# Crystal Research and Technology

*Journal of Experimental and Industrial Crystallography*

*Zeitschrift für experimentelle und technische Kristallographie*

6

Volume 22
1987
Number 6 June

Akademie-Verlag · Berlin

ISSN 0232-1300 Cryst. Res. Technol., Berlin 22 (1987) 6, 753–866, K103–K111

## Directions to the author

In the journal "Crystal Research and Technology – Kristall und Technik" as a rule the following categories of papers will be printed: original papers, review articles, short notes, and book reviews, preferentially in English. Complete information necessary for type-setting and a "Short Title" should be added.

The manuscript should not exceed 10 pages and be type-written on one side only in the format A4 (21 × 29,7 cm) on white ink-proof paper, with a two-lined distance and a left-hand free edge of 3 cm, and submitted in 2 copies. All sheets of the paper belonging to the contribution are continuously to be numbered.

The head of the manuscript should be arranged in this manner: initials and name of the author; institution, and title of the paper. Then two abstracts should follow, one in the original language and the second in German, English, French, or Russian, respectively.

Formulae, signs and symbols, letters etc. being wanting on the type-writer, should be distinctly written with penholder. O as zero is to be underlined, the capital O is not. Greek letters should be underlined with red ink.

Paragraphs should be marked by indenting of at least 3 touches.

Concerning the marking of the manuscript should be noted:
headings should be underlined, sub-headings not, and indented by 9 touches. The titles of tables are not to be underlined. Other markings in the text will be done by the editorial staff. They mean:

light-red 〰〰〰 = small capitals
blue 〰〰〰 = in italics

light-red | perpendicular straight red line on the left edge = small printing

Tables are to be added on separate sheets of paper at the end of the contribution.

Foot notes should be continuously written on a separate sheet of paper.

Figures (Fig.) – drawings and photographs – should be not more than 8 and submitted (2 copies) in a special envelope: photographs as white high-gloss prints (inscription on a transparent covering sheet) and drawings as distinct sketches. On the back of the submitted figures the reduction scale in tenth (9/10, 8/10, 5/10 e.g.), author's name, short title of the paper, and the number of the figure should be given.

Captions for the figures should be submitted on a separate sheet of paper.

References should be written without numbers in alphabetical order of the names at the end of the contribution. Cited names in the text stand for the references. The references should contain the following items, for example:

for books: RAST, K.: Physikalisch-chemische Rechnungen, Berlin 1958
for journals: KRAUSE, H. N., MÜLLER, K.: Cryst. Res. Technol. 16, 70 (1981)

Add, please, the short title, about 6 keywords, and the academic grades of the authors.
The authors will be sent two galley proofs of review articles and original papers. Proof reading of short notes and book reviews will be done by the editorial staff.

For each of these papers a total of 50 reprints is available free of charge. Requests for additional maximum 200 reprints (against payment) should be given together with the submitted manuscript.

Papers should be directed to the editor's office
DDR-7030 Leipzig, Scharnhorststraße 20,
or to the editor Prof. Dr. Hermann Neels,
DDR-7254 Machern, Zeititzer Weg 25,
or to one of the members of the Advisory Board.

Orders can be sent

- in the GDR:
  to the Postzeitungsvertrieb or to the Akademie-Verlag Berlin, Leipziger Straße 3–4, DDR-1086 Berlin;
- in the other socialist countries: to a book-shop for foreign language literature or to the competent news-distributing agency;
- in the FRG and Berlin (West): to a book-shop or to the wholesale distributing agency Kunst und Wissen, Erich Bieber OHG, Wilhelmstr. 4–6, D-7000 Stuttgart 1;
- in the other Western European countries: to Kunst und Wissen, Erich Bieber GmbH, General Wille-Str. 4, CH-8002 Zürich;
- in USA and Canada to the VCH Publishers, Inc., 303 NW 12th Avenue, Deerfield Beach, FL 33442-1705, USA;
- in other countries; to the international book- and journal-selling trade, to Buchexport, Volkseigener Außenhandelsbetrieb der Postfach 160, DDR, DDR-7010 Leipzig, or to the Akademie-Verlag Berlin, Leipziger Straße 3–4, DDR-1086 Berlin.

Journal „Crystal Research and Technology"

Herausgeber: Im Auftrage des Akademie-Verlages Berlin herausgegeben von Prof. Dr. Hermann Neels.

Verlag: Akademie-Verlag Berlin, Leipziger Str. 3–4, DDR-1086 Berlin; Fernruf: 2236221 oder 2236229; Telex-Nr.: 114420; Bank: Staatsbank der DDR, Berlin, Kto.-Nr. 6836-26-20712.

Chefredakteur: Prof. Dr. Hermann Neels, Zeititzer Weg 25, DDR-7254 Machern; Fernruf: Brandis 3442. Anschrift der Redaktion: Scharnhorststr. 20, DDR-7030 Leipzig; Telefon: 310502.

Veröffentlicht unter der Lizenznummer 1526 des Presseamtes beim Vorsitzenden des Ministerrates der Deutschen Demokratischen Republik.

Gesamtherstellung: VEB Druckerei „Thomas Müntzer", DDR-5820 Bad Langensalza.

Erscheinungsweise: Das Journal „Crystal Research and Technology" erscheint monatlich. Die 12 Hefte eines Jahrganges bilden einen Band. Bezugspreis je Band 480,– DM zuzüglich Versandspesen; Preis je Heft 40,– DM. Der gültige Jahresbezugspreis für die DDR ist der Postzeitungsliste zu entnehmen.

Bestellnummer dieses Heftes: 1082/22/6


AN (EDV) 35830
02800

6

Founded 1966 as
Kristall und Technik

# Crystal Research and Technology

*Journal of Experimental and Industrial Crystallography*



VOLUME 22 · 1987 · NUMBER 6

*Akademie-Verlag · Berlin*

| Cryst. Res. Technol. | 22 | 1987 | 6 | 753–760 |
|---|---|---|---|---|

**Original Papers**

B. Březina, M. Havránková

Institute of Physics, Czechoslovak Academy of Sciences, Prague

# Crystal Growth and Domain Structure of Triglycine Sulphate Doped with Pt(II)

The single crystals of triglycine sulphate doped with bivalent Pt-ions (TGS-Pt) were grown in full-shaped form, i.e. with many growth sectors. The domain structure of individual growth sectors was studied by the liquid crystal method. The growth sectors are mostly in a single domain state but spontaneous polarization is reversed in growth sector boundaries, which coincide with charged domain walls. Only the growth sector 001 reveals a block-shaped domain structure, however it differs from usual lenticular domains of the undoped TGS. The concentration of Pt in various growth sectors is different. The growth sectors with the higher concentration of Pt than $3 \times 10^{-2}\%$ exhibit the unipolarity stable against thermal schocks.

Triglycinsulfat dotiert mit zweiwertigen Pt-Ionen (TGS-Pt) wurden als vollausgebildete Kristalle mit vielen Wachstumssektoren gezüchtet. Die Domänenstruktur der individullen Wachstumssektoren wurde mit der Flüssigkristallmethode studiert. Die Wachstumssektoren befinden sich meistens im Eindomänenzustand, aber die Richtung der spontanen Polarisation in der Sektorengrenze ist umgekehrt, wenn sie identisch mit der aufgeladenen Domänenwand ist. Nur im Wachstumssektor 001 findet man eine blockförmige Domänenstruktur, die sich von der gewöhnlichen linsenförmigen Domänenstruktur des undotierten TGS unterscheidet. Die Konzentration von Pt ist in verschiedenen Wachstumssektoren unterschiedlich. Die Wachstumssektoren mit höherer Konzentration als $3 \times 10^{-2}\%$ Pt weisen eine stabile Unipolarität gegenüber thermischem Schock auf.

## 1. Introduction

TGS is monoclinic with the space groups $P2_1/m$ above the Curie temperature $T_c = 49$ °C and $P2_1$ below $T_c$. In the ferroelectric phase, below $T_c$, two equivalent directions of the spontaneous polarization $P_s$ along the polar axis $b$ may occur. Each volume of uniform polarization represents a domain. The adjacent domains have antiparallel $P_s$.

The domains of the undoped TGS are usually in the shape of deformed tubes with a lenticular cross section (Fig. 1). The tubes are oriented along the $b$-axis. The long axes of the lenticular cross sections are situated nearly perpendicular to the $c$-axis. A number of domain boundaries separating domains, referred to as domain walls, depends on many factors. A domain wall density of as-grown crystals staying a few weeks in an electrically nonconductive environment is about 250 per cm².

Březina and Havránková (1985a) have reported the first information on properties of the TGS-Pt crystals. The charged Pt-glycine complexes extremely lower the number of domain walls, often provoking the unipolarity of whole growth sectors. Here, the domain structure of the TGS-Pt crystal was studied in all growth sectors.

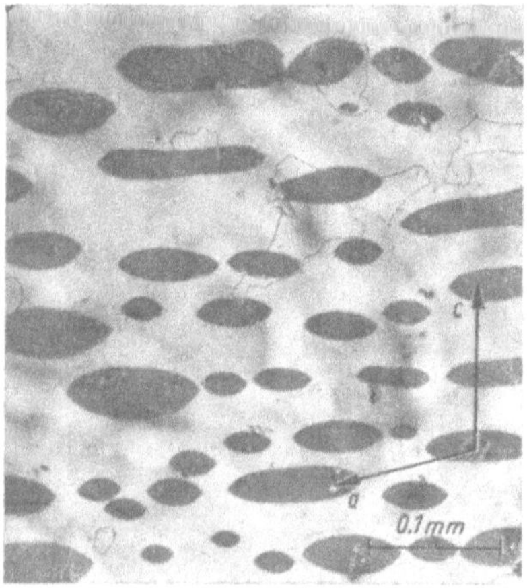


Fig. 1. Lenticular domain structure on the $b$-cleavage surface of the undoped TGS

## 2. Crystal growth and morphology

The single crystal of TGS-Pt was grown in ferroelectric phase by cooling of saturated aqueous solution from 46 to 34 °C. In the growth solution there was 1.5 wt% (related to TGS) of disodiumplatinumtetrachloride. The crystal was grown on a Pt-doped stick seed with the main growth faces {110} and with the longest dimension 6 cm along the $c$-axis (Fig. 2). Both terminal parts of the seed were cut perpendicular to the $c$-axis. Because the seed was fixed in the growth solution on a propeller in the axially symmetrical position along the $c$-axis and a reversible rotation of the seed (30 r.p.m.) was applied, one would expect that only centrosymmetrical hydrodynamic flows could influence the growth morphology. Nevertheless, the crystal habitus reveals a weak asymmetry (Fig. 3) along the pseudomirror plane m (perpendicular to $b$) due to the developed growth sector $0\bar{1}0$ and sporadically developed sector 010[1]). It was found that growth rate ratios of the undoped TGS are equal to[2]) $v_{\langle 100\rangle}:v_{[010]}:v_{[0\bar{1}0]}$:

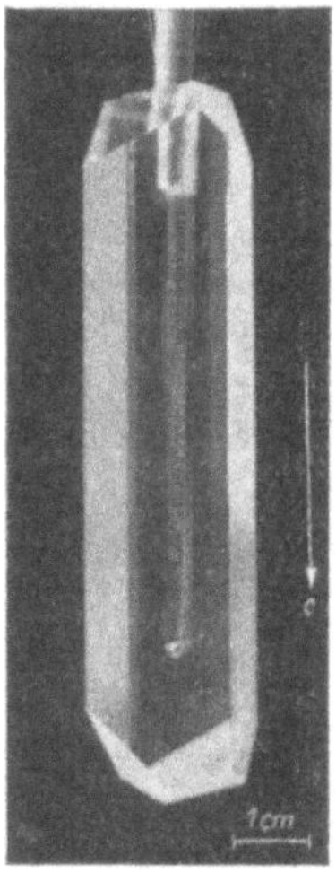


Fig. 2. TGS-Pt crystal in the growth position. The seed is revealed by a light scattering in the regeneration zone

---

[1]) In this paper the individual growth sectors are labelled by Miller indexes of the corresponding growth faces where the parantheses are omitted.

[2]) $v_{\langle 100\rangle} = v_{+a} = v_{-a}$, $v_{\langle 001\rangle} = v_{+c} = v_{-c}$.

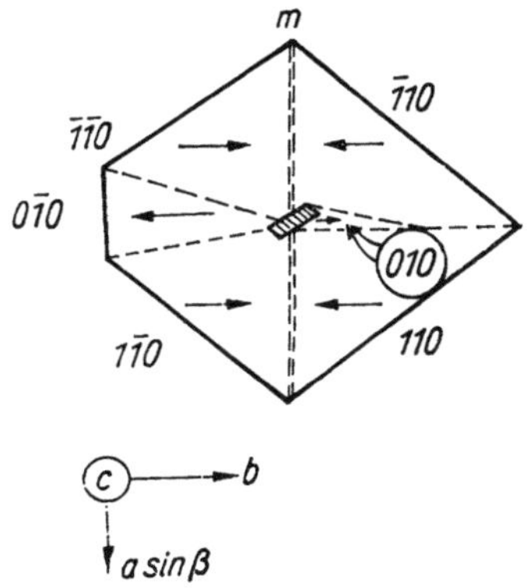


Fig. 3. Schematic drawing of the cut from the middle part TGS-Pt crystal. Dashed lines – growth sector boundaries, arrows – directions of $+P_s$, m – a pseudomirror plane. The orientation after BŘEZINA, HAVRÁNKOVÁ (1985 b)

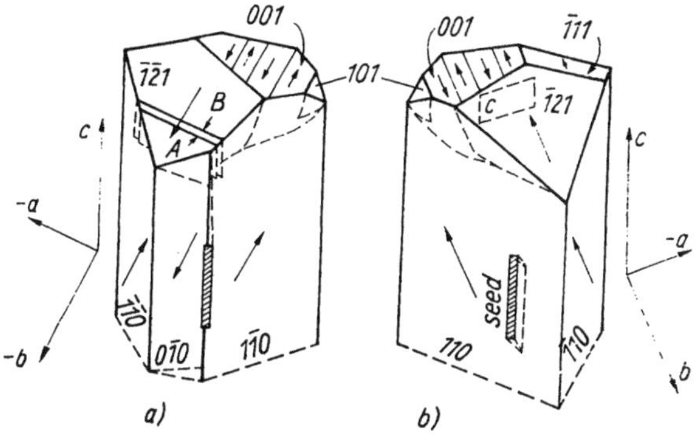


Fig. 4 a, b. Two illustrations of a half of TGS·Pt crystal with the terminal growth faces. Dashed lines – growth sector boundaries, full and dashed lines – domain walls, arrows – directions of $+P_s$. A, B and C – the samples cleaved normal to the $b$-axis

$:v_{\langle 001\rangle}:v_{[110]} = 9.9:10:10:1.3:8.4$ and those ones of the TGS-Pt crystal equal $7.9:10:8.2:13.8:6.8$. The growth rate along $\langle 001\rangle$ of TGS-Pt is ten times higher than in the undoped TGS. In Figures 4 a and b there is one terminal part of our crystal. The orientation according to BŘEZINA and HAVRÁNKOVÁ 1985 b is used. The main growth faces $\{110\}$ were already present on the seed. They create the largest growth sectors. The growth sectors $0\bar{1}0$ and 010 were developing on the seed immediately after the beginning of the growth. Later the 010 sector has disappeared, to the contrary of the $0\bar{1}0$ sector (Fig. 3). At the beginning the crystal recovers the point group symmetry 2/m of the paraelectric phase, but vanishing of the growth face (010) breaks the symmetry to the point group 2 of the ferroelectric phase. The terminal growth faces $\{\bar{1}21\}$, $\{\bar{1}\bar{2}1\}$, $\{001\}$, $\{101\}$ and $\{\bar{1}11\}$ had not been developed on the seed and arised much later during the growth. This is the reason why the corresponding growth sectors are not fixed on the top of the seed (Fig. 4 a, b).

### 3. Observation of domains

Domain patterns were studied on the $b$-plates from various growth sectors. The liquid crystal method (FURUHATA, TORIYAMA; GLOGAROVÁ; NAKATANI, HIROTA) was used for visualizing of domains. On the cleavage surface of ferroelectric TGS or TGS-Pt nematic p-methoxybenzyliden-p-butyl-aniline (MBBA, alfa product) aligns in different ways on domains with opposite orientation of $P_s$. Therefore, samples were covered with MBBA and then with a glass plate with the easy direction on its surface prepared by rubbing on valvet. If the easy direction was parallel to the $c$-axis of the sample, then in the extinction position between crossed polarizers the positive domains (with $+P_s$ on the observed surface) reveal as the bright areas. The negative domains are dark. For parallel polarizers the contrast is opposite.

## 4. Results

Here we introduce the results from the observation of a domain structure in one representative single crystal of TGS-Pt.

From Figure 3 it follows that most of the crystal volume is created by the growth faces {110} with $P_s$ directed inwards the crystal. The boundaries between the opposite polarized growth sectors $\bar{1}\bar{1}0$—$\bar{1}10$ and $1\bar{1}0$—$110$ are not smooth. They are separated in a thickness of about 1 mm by the wedge-shaped lamellars, as if two combs are connected with their spikes. This lamellar domain structure on the $b$-cut is shown in

Fig. 5. Lamellar domain structure in the growth sector boundary $\bar{1}\bar{1}0 - \bar{1}10$ on the $+b$ cleavage surface

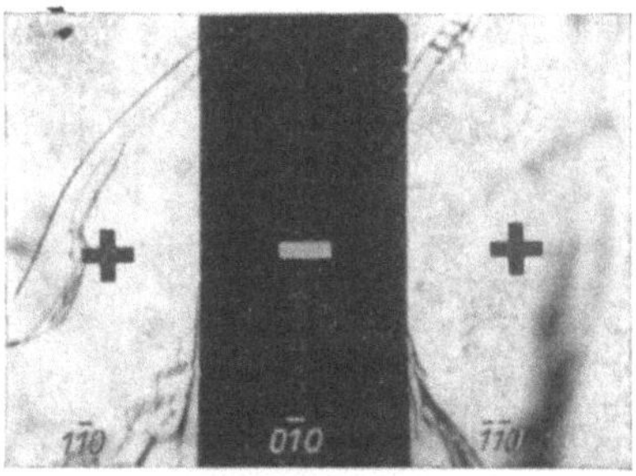


Fig. 6. Domain structure of the growth sectors $1\bar{1}0$, $0\bar{1}0$ and $\bar{1}\bar{1}0$ on the $+b$ cleavage surface (crossed polarizers). Positive domains +, a negative domain −. The irregular lines are cleavage steps

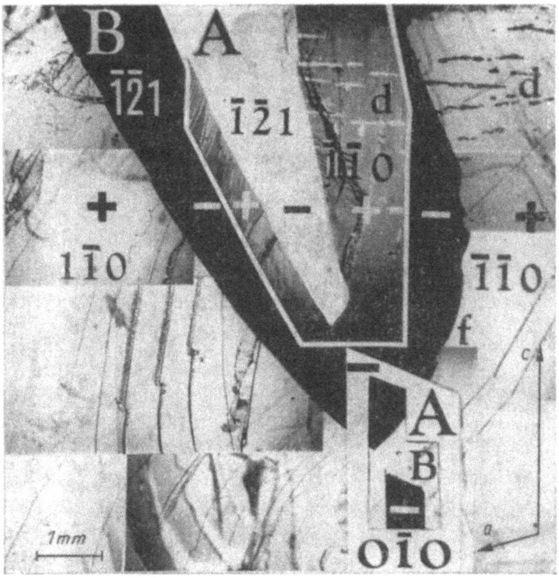


Fig. 7. Photomontage of the domain structure in the samples A and B penetrating the growth sectors $\bar{1}\bar{2}1$, $\bar{1}\bar{1}0$, $0\bar{1}0$ and $1\bar{1}0$ (see Fig. 4 a). Positive domains +, negative domains −, d – minute negative domains; f – domain frittering. The irregular lines are the cleavage steps

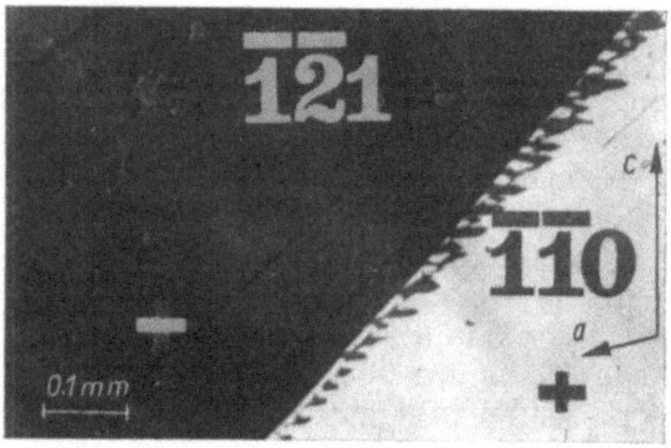


Fig. 8. Detail of the domain frittering from Figure 7

Fig. 9a. Domains of the growth sector 001 on the $-b$ cleavage surface (crossed polarizers) with walls nearly parallel to $(\bar{3}02)$. The bright curved lines are disclinations in the liquid MBBA

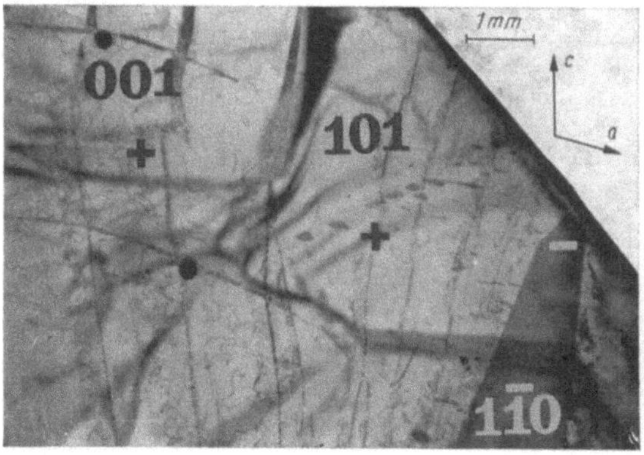


Fig. 9b. Domains of the growth sector 001, 101 and $1\bar{1}0$ on the $-b$ cleavage surface (crossed polarizers). The lenticular domains in 101 sector

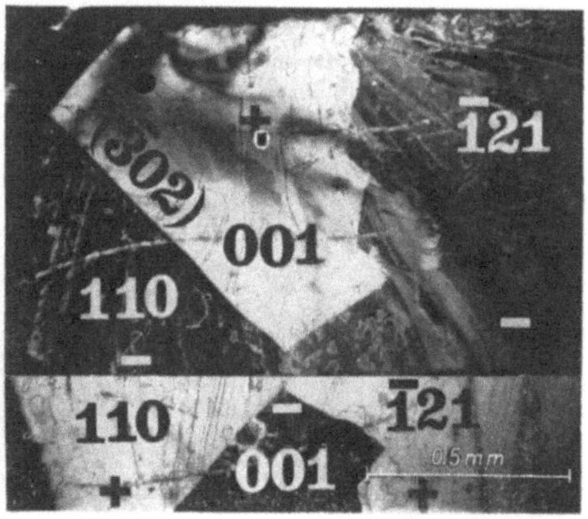


Fig. 9c. Above: Single domains of the growth sectors 110, 001 and $\bar{1}21$ on the $+b$ cleavage surface (crossed polarizers) of the sample C (see Fig. 4b). Below: Matching domain structure on the $-b$ cleavage (crossed polarizers)

Figure 5. The growth sector 0$\bar{1}$0 in Figure 3 reveals an antiparallel $P_s$ with respect to $P_s$ in the neighbouring sectors $\bar{1}$10 and 1$\bar{1}$0. The boundaries 110—0$\bar{1}$0—$\bar{1}\bar{1}$0 are smooth as shown in Figure 6. The growth sector 010 containing at the beginning one single domain with $P_s$ along [010] has disappeared during the growth. All growth sectors shown in Figure 3 are in the single domain state.

Also the terminal growth sectors (Fig. 4) were predominantly in the single domain state, with an exception of the growth sector 001. Therefore, a majority of growth sector boundaries is identical with domain walls. Only in the growth sector 001 domain walls appear which are not identical with growth sector boundaries.

A few examples of the domain structure in terminal growth sectors are in Figure 7 to 9. The domain structure of two samples, A and B, (schematically shown in Figure 4a) is presented in Figure 7. The samples were cleaved across the growth sectors $\bar{1}\bar{2}$1, $\bar{1}\bar{1}$0, 0$\bar{1}$0, 1$\bar{1}$0. A photomontage of intersection of both samples with revealed domains (Fig. 7) was carried out by following procedure. The domains of the sample A were observed on the $+b$ cleavage surface (Fig. 4a) in parallel polarizers and the domains of the sample B on the same $+b$ cleavage surface in crossed polarizers. The negative single domains were revealed on the areas A-$\bar{1}\bar{2}$1 (bright), B-$\bar{1}\bar{2}$1 (dark), A-0$\bar{1}$0 (bright) and B-0$\bar{1}$0 (dark) (Fig. 7). A group of minute negative domains in the vicinity of the growth sector boundary $\bar{1}\bar{2}$1—$\bar{1}\bar{1}$0 (the localities d) is repeated in both samples. At the curved part of $\bar{1}\bar{2}$1—$\bar{1}\bar{1}$0 boundary in the sample B a domain frittering was observed (the locality f). The other domains in Figure 7 are positive. The detailed shot of the domain frittering is shown in Figure 8.

In Figures 9 a, b and c there is the domain structure of the growth sector 001 and its neighbourhood. In Figure 9a large blocks of antiparallel domains a few millimeters thick were observed. Domain walls are predominantly situated parallel with ($\bar{3}$02). In Figure 9b there is the domain structure of a part of the growth sectors 001, 101 and 1$\bar{1}$0. A few minute negative domains with the lenticular form were observed in the growth sector 101 with very low concentration of the dopant (less than $10^{-3}$ wt% of Pt). In Figure 9c there is the domain structure on the $+b$ cleavage surface of the sample C located in Figure 4b. The domain boundary 110—001 is planar and parallel with ($\bar{3}$02). The boundary 001—$\bar{1}$21 is deformed, lying partly along ($\bar{3}$02), partly along non-crystallographical direction. On these parts a fine frittering was observed. The lower part of Figure 9c shows the opposite $-b$ matching surface.

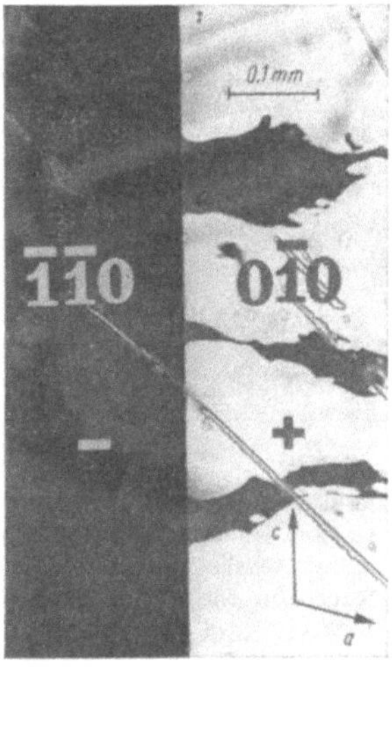


Fig. 10. Domain structure in the neighbouring growth sectors $\bar{1}\bar{1}$0 and 0$\bar{1}$0 after annealing. The $-b$ cleavage (crossed polarizers). New domains arise only in 0$\bar{1}$0 sector with lower concentration of Pt

Further the stability of a single domain state in some growth sectors under the influence of thermal shocks was investigated. The domain structure was studied first of all on virgin samples and then after heating at 80 °C during ten minutes and cooling to 20 °C during five minutes. The unipolarity of the growth sectors 110, $\bar{1}10$, $1\bar{1}0$, $\bar{1}\bar{1}0$ having an average Pt-concentration $3 \cdot 10^{-2}$ wt% was maintained. This concentration corresponds to the occupation of each thousandth elementary cell by one Pt glycine complex. However, in the growth sector $0\bar{1}0$ having by an order lower concentration, $3 \cdot 10^{-3}$ wt%, small antiparallel domains with fringe domain walls have appeared (Fig. 10). The stable unipolarity in the growth sector from growth faces {110} was observed even after 40 hours at 80 °C.

## 5. Discussion

The domain structure of an TGS crystal doped with platinum differs significantly from the domain structure of an undoped crystal. In TGS-Pt most of the growth sectors are in the single-domain state. On the other hand, in an undoped TGS there is a lot of minute lenticular domains in all growth sectors.

A mechanism of the growth sector unipolarity in TGS-Pt differs from that in ATGS (i.e. doped with L, alanine (Keve et al.; Březina et al. 1985c) or D, alanine (Březina et al. 1985b)). In ATGS only one $P_s$ direction in all growth sectors is possible (in LATGS along $[0\bar{1}0]$, in DATGS along [010]). This is due to only one possible position in which the chiral molecule of alanine can be introduced into the lattice of TGS. The position of polar $NH_3^+$ group of L, alanine is in the opposite mirror position compared with D, alanine. It is evident from our experiment that spontaneous polarization $P_s$ in TGS-Pt can be directed in both possible directions along the polar $b$-axis and it is oriented in the majority of the crystal volume into the bulk of a crystal, e.g. as shown in Figure 3 in the largest growth sectors {110}. This is probably connected with the surface conditions on the growth faces {110} arising at the beginning of the crystallization on the surface of the seed. For the unipolarity of the whole growth sectors are responsible Pt-glycine complexes.

The final domain configuration of ferroelectric crystal corresponds to minimum of total free energy including the energy of the depolarization field arising due to the uncompensated bound charges and energy of domain walls. Such an equilibrium domain structure of the undoped TGS is shown in Figure 1. Here, the orientation of the domain walls along $P_s$ causes that div $\boldsymbol{P}_s = 0$ in domain walls. This is not obeyed in the domain walls of TGS-Pt, which correspond to the growth sector boundaries. These boundaries are charged. Nevertheless, a tendency of keeping div $\boldsymbol{P}$ low is evident in the case of TGS-Pt as well. For example, to minimize the depolarization energy of the "head to head" configuration of $P_s$ in the boundaries $110-1\bar{1}0$ and $\bar{1}10-\bar{1}\bar{1}0$ (Fig. 3) they divide into many minute antiparallel domains (Fig. 5).

Well developed block-like domain structure of TGS-Pt in a growth sector bulk was observed only in the growth sector 001 (Figs. 4 and 9a). Here, the domain walls along $(\bar{3}02)$ were formed during the growth by the edge of the growth faces (101) and (001) parallel with $P_s$ and in this case in the wall div $\boldsymbol{P} = 0$.

In Figures 7 and 8 a domain structure frittering is shown, already known in TGS doped with paramagnetic ions or irradiated with $\gamma$-rays (Hilczer et al.). The domain boundary $\bar{1}21-\bar{1}\bar{1}0$ is decorated with minute domains on the curved part only (see the locality f in Fig. 7). This boundary part does not correspond to any crystallographical plane. No frittering was observed on the plane part of the boundary (the upper part) situated along $(\bar{4}01)$. A similar phenomena were observed in Figure 9c on the boundary $001 - \bar{1}21$, where the boundary parts along $(\bar{3}02)$ are not frittered but the curved parts reveal a fine frittering.

A concentration distribution of the Pt-dopant in a full-shaped TGS-Pt crystal is not uniform. It is uniform in bulks of the growth sectors only, but in the growth sector boundaries there exist concentration gradients.

The authors sincerely thank Dr. V. Janovec for valuable discussions, Dr. J. Novák for chemical analyses and Mrs. N. Dvořáková for technical assistance.

### References

BŘEZINA, B., HAVRÁNKOVÁ, M.: Ferroelectrics Letters **4**, 81 (1985) a
BŘEZINA, B., HAVRÁNKOVÁ, M.: Cryst. Res. Technol. **20**, 781 (1985) b
BŘEZINA, B., HAVRÁNKOVÁ, M.: Cryst. Res. Technol. **20**, 787 (1985) c
FURUHATA, Y., TORIYAMA, K.: Appl. Phys. Lett. **23**, 361 (1973)
GLOGAROVÁ, M.: J. Physique **42**, 1569 (1981)
HILCZER, B., PAWLACZYK, Cz., SZCZEŚNIAK, L., MEYER, K. P., SCHOLZ, R.: Materials Science VII, 427 (1981)
KEVE, E. T., BYE, K. L., WHIPPS, P. W., ANNIS, A. D.: Ferroelectrics, **3**, 39 (1971)
NAKATANI, N., HIROTA, M.: Japan J. Appl. Phys. **20**, 2281 (1981)

(Received November 25, 1986)

*Authors' address:*

Dr. B. BŘEZINA, Dipl. Ing. M. HAVRÁNKOVÁ
Institute of Physics, Czechoslovak Academy of Sciences
Na Slovance 2
180 40 Prague 8, Czechoslovakia

| Cryst. Res. Technol. | 22 | 1987 | 6 | 761–766 |
|---|---|---|---|---|

V. Tuček, S. Holeček, L. Kubíček

Department of Chemical Metallurgy and Corrosion Engineering,
Prague Institute of Chemistry and Technology, Prague, Czechoslovakia

# Natural Convection during Directional Solidification of an Al–Zn Eutectic

A study was made of the buoyancy convection of melt provoked by temperature gradient of an orientation identical to that of the gravity vector. The convection was assessed on the basis of temperature fluctuations in the melt. The critical Rayleigh numbers for periodic and turbulent flow were determined. Both periodic and turbulent flow gives rise to transverse "bands" in the structure of the eutection obtained by directional solidification. The intensity of unstable flow affects the number of eutectic crystallites in the sample.

В рабоме исследованна гравитационная конвекция вызана гррдиентом темпера-тры ориентированным согласно с гравитацией. Конвекция была оценивана с помощью флюктуаций температуры в расплаве. Были установлены критические величины числа Rayleigh для возникновения периодического и турбулентного движения расплава. Нестабильное движение в расплаве при направленной кристаллизации способствует возникновению поперечных "полосы" в структуре эвтектики. Интенсивность нестабильного движения расплава влияет на число эвтектических кристаллов в образце.

## 1. Introduction

One of the trends in current materials research is the studies of the effects of the force of gravity on the crystallization of metals. The most significant phenomenon by which the conditions of crystallization in outer space differ from the terrestrial conditions is the considerable limitation, or even exclusion, of natural convection of the melt due to differences in density (Steg).

Studies into the mechanism of flow are rather difficult in non-transparent metallic liquids, inasmuch as the application of optical techniques is precluded. The technique which is used most frequently for this purpose is the measurements of temperature gradients and temperature fluctuations. Comparisons of the results obtained with the mechanics of flow of transparent liquids are used to draw conclusions as to the behavior of the non-transparent melts of metals and semiconductors (Müller, Neumann). Depending on the character of the temperature fluctuations, the types of flow which can be distinguished are laminar flow free from fluctuations, periodic flow where the temperature fluctuations are periodic, and turbulent where the temperature fluctuations are aperiodic.

The hydrodynamic state of liquid enclosed in a sealed vessel can be characterized by a combination of dimensionless numbers (Ostrach) using the
Rayleigh number

$$\mathrm{Ra} = \frac{\beta \cdot g \cdot G_t \cdot H^4}{a \cdot \nu}$$

Prandtl number

$$\mathrm{Pr} = \frac{\nu}{a}$$

and the ratio of the vessel dimensions, $H/d$, where

$g$ the acceleration due to gravity
$\beta$ the coefficient of volumetric thermal dilation
$\nu$ the kinematic viscosity
$a$ the temperature conductivity
$G_t$ the average temperature gradient in melt (calculated from temperature difference between bottom and top level)
$H$ the height of liquid
$d$ the diameter of the cylindrical vessel

For prediction of the character of flow in the sample, one has to know the critical values of the Rayleigh number which determine the conditions for the switchovers from one to another type of flow. The critical value $\mathrm{Ra}^{c1}$ determines the conditions under which the laminar flow will prevail in the system; the value $\mathrm{Ra}^{c2}$ sets the conditions for a transition between laminar and periodic flow; and the value $\mathrm{Ra}^{c3}$ is the boundary between periodic and turbulent flow. Inasmuch as the friction of liquid on the vessel valls affects the character of flow, the critical values of the Rayleigh number depended on the $H/d$ ratio. The critical value of $\mathrm{Ra}^{c1}$ is the same for liquids having either low high Prandtl numbers (Müller et al.), whereas the critical value of $\mathrm{Ra}^{c3}$ is rather different for liquids having low Prandtl numbers (metallic melts have $\mathrm{Pr} \approx 2 \cdot 10^{-2}$) and high Prandtl numbers (water $\mathrm{Pr} \approx 6.7$).

This paper presents the results of a study of the natural convection within an Al–Zn eutectic melt during directional solidification. The work was aimed at determining the critical values of the Rayleigh number in the Al–Zn eutectic melt, and at demonstrating the effects of the individual kinds of flow involved upon the structure of the oriented, solidified eutectic.

## 2. Experimnetal

Directional solidification of the Al–Zn eutectic was conducted using a modified Bridgman Stockbarger method. The Al–Zn alloy of eutectic composition (94.9% Zn) was enclosed in a graphite ampoule of cylindrical shape having 11 mm internal diameter and 90 mm length. The top end of the ampoule was fixed to a copper cooler with circulating water. A resistance furnace was slid onto the sample thus prepared. After melting and after temperature stabilization, the furnace started moving in the direction of the force of gravity, at a rate of $1.13\ \mu\mathrm{m} \cdot \mathrm{s}^{-1}$. The sense of the temperature gradient in the sample coincided with that of gravity, so that the melt was virogously agitated at the beginning of the experiment. In the commonly used arrangement with bottom cooling of the ampoule, a stable situation develops where the melt is not agitated.

By varying the furnace temperature and the cooling water parameters, an average starting temperature gradient in the melt was set at 8, 20, or $30\ \mathrm{K} \cdot \mathrm{cm}^{-1}$.

In order to monitor the advance of the crystallization front and to measure temperatures within the sample, a corundum capillary of 1.7 mm outer diameter was placed in the axis of the graphite ampoule. Up to three thermocouples Ni–NiCr of 0.2 mm diameter could be packed in the capillary. This bundle of thermocouples could be shifted in and out at will within the capillary during the run. Thus we were able to record temperature fluctuations and to measure the height of the melt and the temperature difference within the melt.

The ampoules used for samples to be used for studies of structure only held one single Ni–NiCr thermocouple, position within the ampoule axis, at a spacing of 10 mm from the hotter end. After directional solidifcation the samples were sliced longitudinally and one half of the sample was also sliced transversally to 10 mm layers. The longitudinal as well as the transverse sections were ground and polished and were etched with 0.5% HF. The sections were used for structural examinations.

## 3. Results

The studies of the temperature field established within entirely molted samples made it clear that the highest gradients can be encountered between the 2nd and 3rd centimeters, counted from the cool end of the sample (Tab. 1). The average temperature gradients were gradually decreasing as the crystallization front advanced though the sample. The reduction of the average gradients of temperature in the sample ($G_t$) can be described as function of melt height by the following equations:

at 30 K · cm$^{-1}$ initial gradient $$G_t = \frac{270}{18 - H},$$

at 20 K · cm$^{-1}$ initial gradient $$G_t = \frac{200}{19 - H},$$

at 8 K · cm$^{-1}$ initial gradient $$G_t = \frac{250}{40 - H}.$$

The average advance rates of the crystallization front were about twice as high as the furnace travelling rate which was 1.13 μm · s$^{-1}$. The minimum front advance rates were encountered between the 2nd and 3rd centimeters, thus corresponding to the maximum temperature gradients in the sample (cf. Tab. 1).

Table 1
Temperature gradients and crystallization front advance rates in directional solidification

| | | | |
|---|---|---|---|
| average initial temperature gradient, K · cm$^{-1}$ | 8 | 20 | 30 |
| maximum temperature gradient, K · cm$^{-1}$ | 11 | 23 | 33 |
| average rate of advance of the crystallization front, μm · s$^{-1}$ | 2.1 | 1.1 | 1.2 |
| minimum rate of advance of the crystallization front, μm · s$^{-1}$ | 1.3 | 0.8 | 0.8 |

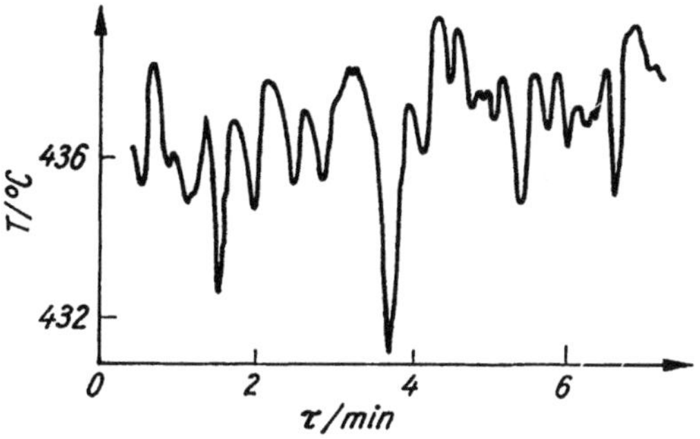


Fig. 1. Irregular temperature fluctuations during crystallization at a temperature gradient 8 K · cm$^{-1}$

Table 2
Parameters of the temperature fluctuation functions

| | | | |
|---|---|---|---|
| Average initial temperature gradient, K · $cm^{-1}$ | 8 | 20 | 30 |
| Amplitude of aperiodic fluctuations, K | 4 | 11 | 16 |
| Height of melt $H_3$, cm | 6.5 | 4.1 | 3.6 |
| Amplitude of periodic fluctuations, K | 1.2 | 1.4 | 1.6 |
| Frequency of fluctuations, $s^{-1}$ | 0.048 | 0.066 | 0.085 |
| Height of melt $H_2$, cm | 3.8 | 2.8 | 2.5 |

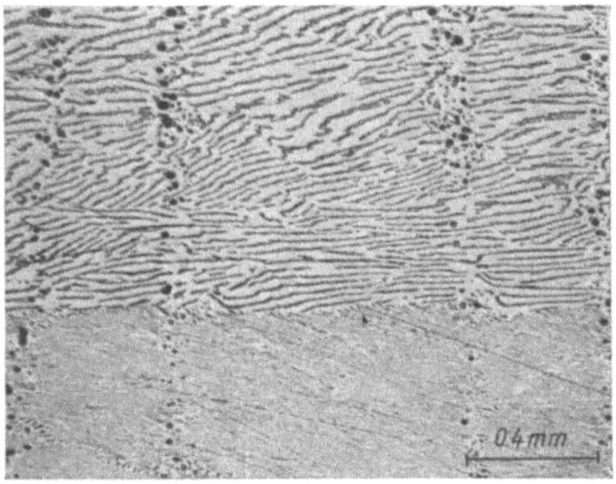


Fig. 2. Irregular "bands" occurring transversely in the structure of an eutectic solidified at a temperature gradient of 8 K · $cm^{-1}$

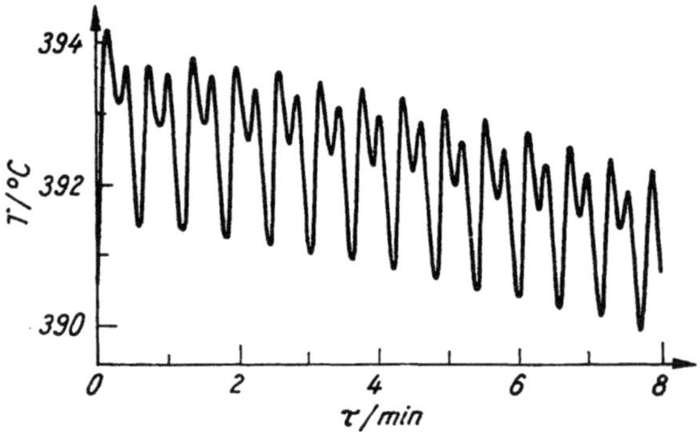


Fig. 3. Periodic temperature fluctuations during crystallization at a temperature gradient of 8 K · $cm^{-1}$

In fully molten samples, evidence was found of irregular temperature fluctuations (Fig. 1). The amplitude of fluctuations grew higher with increasing temperature gradient (see Tab. 2). During directional crystallization, the level (height) of the melt in the ampoule kept decreasing. The irregular temperature fluctuations were converted to periodic fluctuations at the melt heights $H_3$ (cf. Tab. 2, Fig. 3). The amplitudes and frequencies of these fluctuations at the centerpoint of the melt are shown in Table 2. The fluctuation amplitude was changing with the position of the thermocouple, whereas the frequency remained the same. As the height of melt was pushed down further the temperature fluctuations faded away. The heights of melt at which the temperature fluctuations ceased ($H_2$) are also listed in Table 2.

As demonstrated by the studies of structure, those parts of the samples which solidified in the presence of irregular temperature fluctuations contained irregularly

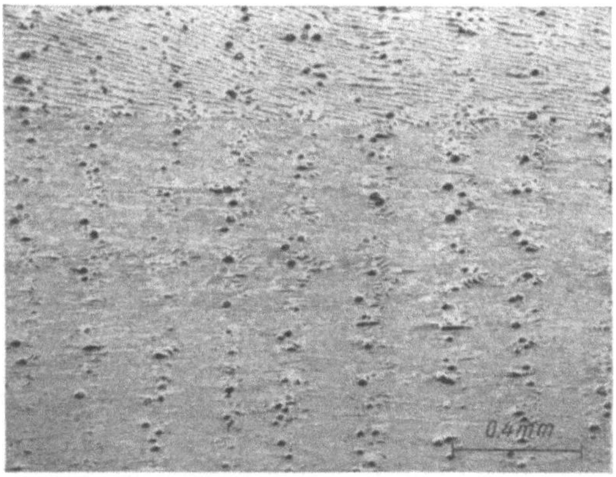


Fig. 4. Periodically arranged transverse bands in the eutectic structure solidified at a temperature gradient of 8 K · cm⁻¹

arranged transverse bands denoting accumulations of lamellar defects (Fig. 2). Those parts of the samples which grew in the presence of periodic temperature fluctuations in the melt contained periodically arranged bands (Fig. 4). The bands were no longer observed after the melt height dropped to a level corresponding to the decay of the temperature fluctuations. In samples that grew at higher temperature gradients the structure contained a higher number of eutectic crystallites than that of samples that grew under conditions of the lower gradients.

## 4. Discussion

The switchover from aperiodic to periodic temperature fluctuations and the decay of the fluctuations, as well as the corresponding changes in the character of the transverse bands encountered in the structure, are caused by changes in the character of flow of the liquid during an directional solidification of the eutectic melt. For this reason, the aforementioned measuring technique could be employed to establish the critical values of the Rayleigh number, $Ra^{c2}$ and $Ra^{c3}$.

During the experiment, the melt height and the temperature gradient were decreasing, and so was the value of the Rayleigh number for the system. Values from Table 3 were used to calculate the Rayleigh numbers. At the melt height $H_3$ the value of $Ra^{c3}$ was overstepped, and at the height $H_2$ the value of $Ra^{c2}$ was exceeded.

Inasmuch as three different temperature gradients were used in the measurements, it was possible to determine three pairs of values $Ra^{c2}$ and $Ra^{c3}$ (Fig. 5). The critical values $Ra^{c1}$ could not be determined by the technique used here. Figure 5 indicates a values determined by CATTON. The values of $Ra^{c2}$ and $Ra^{c3}$ increase monotonously over the $H/d$ interval studied. $Ra^{c2}$ has risen from $9.7 \cdot 10^4$ at $H/d = 2.3$ to $2.0 \cdot 10^5$ at $H/d = 3.5$; $Ra^{c3}$ from $4.7 \cdot 10^5$ at $H/d = 3.3$ to $1.8 \cdot 10^6$ at $H/d = 5.9$. The ascertained $Ra^{c2}$ and $Ra^{c3}$ values correspond with values obtained for a gallium melt (MÜLLER et al.). For a more accurate determination of the critical Rayleigh number it would be necessary to use an ampoule with perfectly insulated walls, so as to elimi-

Table 3
Physical properties of Al–Zn eutectic melt

| | |
|---|---|
| $a/cm^2 \cdot s^{-1}$ | 0.22 |
| $\nu/cm^2 \cdot s^{-1}$ | $4.1 \cdot 10^{-4}$ |
| $\beta/K^{-1}$ | $1.25 \cdot 10^{-4}$ |

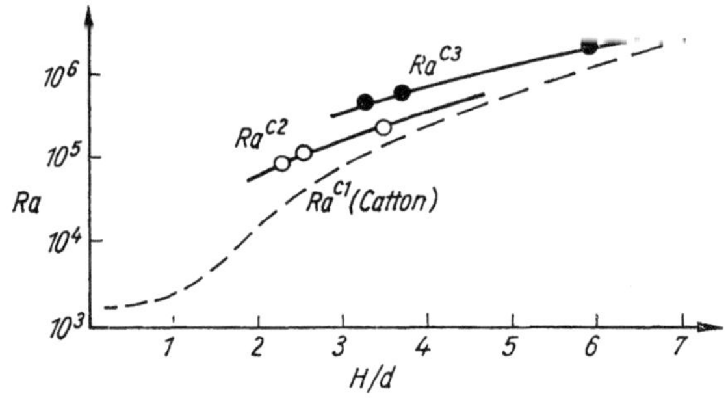


Fig. 5. Critical values of the Rayleigh number as function of the $H/d$ ratio

nate the effect of the radial temperature gradient in the sample; this effect was disregarded in this study.

The temperature fluctuations in the melt during directional crystallization are responsible for the fluctuations in the rate of advance of the crystallization front. These irregularities are the cause of the pileups of defects during lamellar growth, of the type which suggests in appearance a transverse interruption of the lamellar structure in the sample obtained by directional solidification.

As the temperature gradient grows in the sample and, hence, as the Rayleigh number increases, there is a more vigorous agitation of the melt, with the consequence of greater temperature fluctuations. Even a local overcooling may occur which is necessary for solid phase nuclei to form in the melt ahead of the crystallization front. Apparently, the resultant structure contains therefore a greater amount of fine eutectic crystals.

## 5. Conclusions

The critical values of the Rayleigh number were calculated for an Al–Zn eutectic melt which are characteristic of the occurrence of periodic and turbulent flow. These value are lower than those characterizing the occurrence of unstable flow in water (Müller et al.).

The periodic and turbulent flow within the melt causes temperature fluctuations. In these case of directional crystallization, this brings about the formation of transverse interruptions of the lamellar structure and an increase in the quantity of eutectic crystallites in the sample.

### References

Catton, I.: AICHE Journal **16**, 594 (1970)

Müller, G., Neumann, G.: Proc. 4th European Symp. on Materials Sciences under Microgravity, Madrid 1983, p. 285

Müller, G., Neumann, G., Weber, W.: J. Cryst. Growth **70**, 78 (1984)

Ostrach, S.: Proc. 2nd European Symp. on Materials Sciences in Space, Frascati 1976, p. 41

Steg, O.: Kosmicheskaya technologiya, Moscow 1980

*Authors' address:*

V. Tuček, Dozent Dr. Ing. S. Holeček, L. Kubíček
Department of Chemical Metallurgy and Corrosion Engineering,
Prague Institute of Chemistry and Technology
Suchbatarova 5
16628 Prague 6
Czechoslovakia

Cryst. Res. Technol. | 22 | 1987 | 6 | 767–776

H. KÜHNE

Akademie der Wissenschaften der DDR
Institut für Halbleiterphysik, Frankfurt (Oder)

# The Current Understanding of Epitaxial CVD Silicon Doping in the Light of Modelling and Theory Development (III)

## The Law of Epitaxial Layer Doping and its Limitation

Various doping mechanisms discussed in recent literature are applied to the theoretical concept derived in the previously published Part II of the paper, whereby a variety of hypothetical laws of dopant incorporation is revealed. Those laws cannot be distinguished from each other by doping experiments, but are testable, at least in principle. When substituting three doping-process-relevant empirical constants, all of them might be changed into that single empirical, but theory-aided model equation, which enables apitaxial layer doping to be described as a result of intended variations of the main process parameters: partial pressure of the dopant source, layer growth rate, total pressure, and deposition temperature.

Gegenwärtig in der Literatur diskutierte Dotierungsmechanismen werden mit dem im bereits veröffentlichten Teil II der Arbeit entwickelten theoretischen Konzept verknüpft und eine Reihe hypothetischer Dotandeneinbaugesetze abgeleitet. Diese Gesetze lassen sich durch die Auswertung von Dotierungsexperimenten allein nicht voneinander unterscheiden, obwohl sie, zumindest prinzipiell, überprüfbar sind. Durch das Einsetzen von drei, aus dem Dotierungsprozeß gewonnenen empirischen Konstanten werden alle diese hypothetischen Gesetze in ein und dieselbe empirische, aber Theorie gestützte Modellgleichung umgewandelt. Sie erlaubt die Beschreibung des Dotierungsprozesses aus der Wirkung der wesentlichen Prozeßparameter: Partialdruck der Dotandenquellsubstanz, Schichtwachstumsrate, Totaldruck und Schichtabscheidungstemperatur.

## 1. Introduction

In the previously published part I of the paper a brief description of silicon epitaxy was given and the main process controlling parameters: input pressure of the dopant source ($p^0_{\text{dope}}$), layer growth rate ($v_{\text{epi}}$), total pressure ($p_{\text{tot}}$), and deposition temperature ($T_{\text{epi}}$) were introduced. Considering the doping process analogous to signal transfer in cybernetic systems an empirically based control equation of layer doping was formulated (equ. 9/I)

$$N(x) = \frac{A \cdot p^0_{\text{dope}}}{(1 + B \cdot v_{\text{epi}})} \cdot \left[1 - \exp - \frac{\Delta x}{L}\right]. \qquad (9/\text{I})^{1)}$$

Equ. (9/I) describes both steady state and non-steady state dopant incorporation (REIF et al.). Unfortunately the effect of total pressure on layer doping is not expressed by equ. (9/I).

[1]) The Roman numbers I and II refer to the previously published irrespective parts of the paper.

In the also previously published Part II of the paper theoretical foundations selected from related fields of science (i.e. thermodynamic and chemical equilibria, kinetics of surface-catalized reactions, open systems and steady state equilibria, solid state physics of semiconductors) were applied to explain the doping process. Taking into account not only the decomposition of the dopant source by heat (1/II),

$$p^0_{AsH_3} = \sum_1^n i \cdot p^{eq}_{As_iH_j} \approx p^{eq}_{AsH_2}|_{p_{tot}=1\,atm} \approx p^{eq}_{AsH_2} + p^{eq}_{As}|_{p_{tot}\ll 1\,atm}, \qquad (1/II)^{1)}$$

where: $1 \leqq i \leqq 4$, $0 \leqq j \leqq 4$
but also the incomplete ionisation of incorporated dopant atoms at deposition temperatures (15/II),

$$(N_{As_{Si}})_{T_{epi}} = F_1 \cdot (N_{As^+_{Si}})_{tot}|_{N_e = N_i} \qquad (15/II)$$

where

$N_e$ Concentration of majority carrier
$N_i$ Concentration of intrinsic carrier

the general differential equation (12/II) was derived by considering the possibility of forward and backward directed chemical reactions to be rate limiting steps in the overall doping reaction.

$$\frac{dI_{dope}}{dt} \cdot \frac{dt}{dx} = \frac{1}{v_{epi}} \cdot [I_1 - (I_2 + I_{dope})] \qquad (12/II)$$

where: $dx/dt = v_{epi}$.
A specific solution of this differential equation is supposed to be the correct and complete law of dopant incorporation.

In Part III of the paper doping mechanisms discussed in more recent literature are examined by applying them to equ. (12/II). A variety of hypothetical incorporation laws are derived and concluded that all these laws be represented by only one model equation having a more general nature. This equation contains three empirical constants which can be obtained by doping experiments and explained by theory, at least in principle.

## 2. The simple conception

One of the first attempts to explain epitaxial layer doping was based on the supposition that only dopant atoms can be transferred from the gas into the silicon layer (SHEPERD). Though, calculations of the equilibrium compositions of such systems soon showed that in comparison to other decomposition products dopant atoms should exist within limits of negligible concentrations only, this idea had not been abandoned (BLOEM et al.). Recently the same basis idea was used to explain lateral autodoping effects of arsenic and phosphorus (GRAEF et al.). In the course followed in the present paper the following reaction steps must be considered:

$$\left.\begin{array}{ll} (As)_g \rightarrow As_{Si}, & I_{10} = k_{10} \cdot p^{eq}_{As}, \\ As_{Si} \rightarrow (As_g), & I_{20} = k_{20} \cdot (N_{As_{Si}})_{T_{epi}} \end{array}\right\} K_{inc} = \frac{k_{10}}{k_{20}} \qquad (1a), (1b)$$

$$I_{dope} = v_{epi} \cdot \frac{(N_{As_{Si}})_{T_{epi}}}{F_1} \qquad (1c)$$

$$As_{Si} \leftrightharpoons As^+_{Si} + e, \qquad K_{ion} = K^*_{ion} \cdot \frac{(N_{As^+_{Si}})_{T_{epi}}}{(N_{As_{Si}})_{T_{epi}}} \qquad (1d)$$

Substituting the equs. (1a) through (1d) into equ. (12/II) leads from:

$$\frac{d(N_{\mathrm{As_{Si}}})_{T_{\mathrm{epi}}}}{dx} = \frac{1}{v_{\mathrm{epi}}}\left[k_{10} \cdot p_{\mathrm{As}}^{\mathrm{eq}'} - k_{20} \cdot \left(1 + \frac{v_{\mathrm{epi}}}{F_1 \cdot k_{20}}\right) \cdot (N_{\mathrm{As_{Si}}})_{T_{\mathrm{epi}}}\right] \tag{2}$$

to the following hypothetical incorporation law:

$$(N_{\mathrm{As_{Si}^+}})_{\mathrm{tot},x} = \frac{K_{\mathrm{inc}} \cdot p_{\mathrm{As}}^{\mathrm{eq}}}{F_1 \cdot \left(1 + \frac{v_{\mathrm{epi}}}{F_1 \cdot k_{20}}\right)} \cdot \left[1 - \exp - \frac{k_{20}}{v_{\mathrm{epi}}} \cdot \left(1 + \frac{v_{\mathrm{epi}}}{F_1 \cdot k_{20}}\right) \cdot \Delta x\right] \tag{3}$$

when $(N_{\mathrm{As_{Si}}})_{T_{\mathrm{epi}}} = 0$ at $t = 0$ and $(N_{\mathrm{As_{Si}}})_{T_{\mathrm{epi}}} = N_\infty$ at $t = \infty$ are chosen as limits of the integration. The structure of the equ. (3) conforms to that of the empirical equation (9/I), what is quite well visible when empirical constants are substituted:

$$(N_{\mathrm{As_{Si}^+}})_{\mathrm{tot},x} = \frac{A \cdot p_{\mathrm{eq}}^{\mathrm{As}}}{(1 + B \cdot v_{\mathrm{epi}})} \cdot \left[1 - \exp - \frac{\Delta x}{L}\right]. \tag{3a}$$

The constants $A$, $B$, and $L$ formerly used in connection with equ. (9/I) are now defined theoretically. Thus $A \equiv K_{\mathrm{inc}}/F_1$ is a thermodynamic constant depending on temperature, $B \equiv (k_{20} \cdot F_1)^{-1}$ constitutes a mixed kinetic-thermodynamic constant also depending on temperature only. Constant $L$ is expressed by an algebraic term given by equ. (4) as a function of layer growth

$$\frac{1}{\tau} = \frac{v_{\mathrm{epi}}}{L} = k_{20} \cdot (1 + B \cdot v_{\mathrm{epi}}) = k_{20} + \frac{v_{\mathrm{epi}}}{F_1} = a + b \cdot v_{\mathrm{epi}}\,. \tag{4}$$

Equ. (4) in its empirical version given with the right-hand term had formerly been deduced by Reif et al. from results of non-steady state doping experiments. Analytical versions have recently been published by Wong et al. and independently by Kühne. As to equ. (3), the effects of such important process parameters as the input pressure of the dopant source and the total pressure are not expressed by equ. (3). In connection with the above hypothetical law of incorporation their influencies may be concluded from an independent function $p_{\mathrm{As}}^{\mathrm{eq}} = f(p_{\mathrm{AsH_3}}^0, p_{\mathrm{tot}} T_{\mathrm{epi}})$ which is a sub-total of the calculation of the equilibrium compositions of the As–H–(Si) system. Fortunately this disadvantage offers a possibility of testing the content of truth of equ. (3). The total pressure dependence of the arsenic concentration in the layer must be equal to the total pressure dependence of the atomic arsenic concentration in the gas at equilibrium conditions. The two dependencies are compared to each other in equ. (5) at two selected levels of total pressure. The experimental data of the arsenic in the layer were originally obtained by Herring and applied in the context of equ. (5) as evaluated by Kühne (c.f. Fig. 4). The data of the calculated equilibrium compositions were taken from Graef et al.

$$\frac{(N_{\mathrm{As_{Si}^+}})_{\mathrm{atm}}}{(N_{\mathrm{As_{Si}^+}})_{0.1\,\mathrm{atm}}} \equiv \left[\frac{(p_{\mathrm{As}}^{\mathrm{eq}})_{\mathrm{atm}}}{(p_{\mathrm{As}}^{\mathrm{eq}})_{0.1\,\mathrm{atm}}}\right]_{\mathrm{cal}} = 3.1 \neq 1.6 = \left[\frac{(N_{\mathrm{As_{Si}^+}})_{\mathrm{atm}}}{(N_{\mathrm{As_{Si}^+}})_{0.1\,\mathrm{atm}}}\right]_{\mathrm{exp}} \tag{5}$$

where: $p_{\mathrm{AsH_3}}^0 = \mathrm{const}$, $T_{\mathrm{epi}} = 1350\ \mathrm{K}$

The inequality of the two quantities compared is evident. Additionally it should be noted that two premises are inherent parts of the negative conclusion drawn above. First the calculation of the equilibrium composition is accepted to give the true result, i.e. the calculated values of the equilibrium pressure of arsenic atoms are considered to conform to the real state, and second the doping experiments and the evaluation of the corresponding data are accepted to have been done without an error.

## 3. The extended conception

As in the section above it is assumed that dopant incorporation be realized by the transfer of dopant atoms from the gas to the layer, but, in contrast, it is suggested that the reproduction of the consumed dopant atoms within the gas is the rate limiting step. Each kind of particles in the gas, containing dopants, decomposes with a rate approachingly proportional to the factor $y_{As_iH_j} \cdot p^0_{AsH_3}$, if, for example, doping with arsenic is considered. The resulting chemical reaction flow stabilizes the concentration of gaseous arsenic atoms. Fortunately, extreme differences between the different factors $y_{As_iH_j}$ allow to neglect the majority of the components except one or two (c.f. equs. (2/II) and (3/II). Such an incorporation mechanism has recently been suggested by HINTZEN et al. to explain certain pecularities of boron incorporation

$$p^0_{AsH_3} \approx p^{eq}_{AsH_2}|_{p_{tot}=1} \approx p^{eq}_{AsH_2} + p^{eq}_{As}|_{p_{tot}\ll 1\,atm} \,. \qquad (2/II, 3/II)$$

At atmospheric pressure, returning to arsenic, the following scheme is needed to fill equ. (12/II) with a new content

$$AsH_2 \rightarrow (As)_g + H_2 \,, \qquad I_{11} = k_{11} \cdot p^{eq}_{AsH_2} \approx k_{11} \cdot p^0_{AsH_3} \,, \qquad (6a)$$

$$H_2 + (As)_g \rightarrow AsH_2 \,, \qquad I_{21} = k_{21} \cdot p^{eq}_{H_2} \cdot p^{eq}_{As} \approx k_{21} \cdot p_{tot} \cdot p^{eq}_{As} \qquad (6b)$$

with: $K_{dec} = k_{11}/k_{21}$

$$I_{dope} = \frac{v_{epi}}{F_1} \cdot (N_{As_{Si}})_{T_{epi}} \,, \qquad (6c)$$

$$(As)_g \leftrightharpoons As_{Si} \,, \qquad K_{inc} = \frac{(N_{As_{Si}})_{T_{epi}}}{p^{eq}_{As}} \,. \qquad (6d)$$

Taking into account the equs. (6a) through (6d) equ. (12/II) must be formulated as follows:

$$\frac{1}{v_{epi}} \cdot \frac{dp^{As}_{eq}}{dt} = \frac{1}{v_{epi}} \cdot \left[k_{11} \cdot p^{eq}_{AsH_2} - k_{21} \cdot \left(p^{eq}_{H_2} + \frac{K_{inc}}{F_1 \cdot k_{21}} \cdot v_{epi}\right) \cdot p^{eq}_{As}\right] . \qquad (7)$$

Integration of equ. (7) leads to the hypothetical law of dopant incorporation given by equ. (8), which fulfills all those expectations that might be associated with such an incorporation law

$$(N_{As^+_{Si}})_{tot,x} = \frac{A \cdot p^{eq}_{AsH_2}}{(p^{eq}_{H_2} + B \cdot v_{epi}} \cdot \left[1 - \exp - \frac{\Delta x}{L}\right] \approx \frac{A \cdot p^0_{AsH_3}}{(p_{tot} + B \cdot v_{epi})} \times$$

$$\times \left[1 - \exp - \frac{\Delta x}{L}\right] . \qquad (8)$$

where: $p_{tot} \approx 1$ atm, $A = \dfrac{K_{dec} \cdot K_{inc}}{F_1}$, $B = \dfrac{K_{inc}}{F_1 \cdot k_{21}}$,

$$\frac{v_{epi}}{L} = k_{21} \cdot p^{eq}_{H_2} + \frac{K_{inc}}{F_1} \cdot v_{epi} = a \cdot p_{tot} + b \cdot v_{epi} \,,$$

and: $A/B = k_{21} \cdot K_{dec}$, $b/B = k_{21}$, $A/b = K_{dec}$, $b/a = B$, $a = k_{21}|_{p_{tot}=1\,atm}$ .

Equ. (8) directly links the dopant concentration in the layer with the main process parameters under consideration. The effect of temperature, of course, is determined by the physico-chemical nature of the theoretical constants. Additionally equ. (8) is in coincidence with the empirical model equation (9/I) as far as its range of validity is limited to atmospheric pressure.

When layer doping is carried out under low pressure, a second kind of particles must be taken into account (c.f. 3/11). The corresponding incorporation equation (9), which should be compared with equs. (5/I) and (6/II), will be given without any further discussion.

$$(N_{As^+_{Si}})_{tot} = \frac{K_{dec} \cdot K_{inc} \cdot p^0_{AsH_3}}{F_1 \cdot \left[p_{tot} + \frac{K_{inc}}{F_1 \cdot k_{21}} \cdot v_{epi} + K_{dec}\right]} \approx \frac{K_{inc} \cdot p^0_{AsH_3}}{F_1 \cdot \left[1 + \frac{p_{tot}}{K_{dec}}\right]} \tag{9}$$

where: $p_{tot} < 1$ atm, and: $\frac{K_{inc}}{F_1 \cdot k_{21}} \cdot v_{epi} \ll 1$.

## 4. The improved conception

Many authors prefer the idea that the particles participating in the dopant transfer from the gas into the silicon layer are intermediately adsorbed on the surface (REIF et al.; WONG et al.; KÜHNE). Adsorption and, corresponding to more recent publications, both adsorption and desorption are suggested as the rate limiting steps. Adsorption should be linked with the chemical decomposition of particles yielding dopant atoms suitable for incorporation

$$* + A^*_sH_2 \rightarrow As^* + H_2\,, \qquad I_{ad} = k_{12} \cdot (N_0 - N^*) \cdot p^{eq}_{AsH_2} \approx$$

$$\approx k_{12} \cdot N_0 \cdot p^0_{AsH_3}\,, \tag{10a}$$

$$A^*_s + H_2 \rightarrow AsH_2 + *\,, \qquad I_{des} = k_{22} \cdot p^{eq}_{H_2} \cdot N^*_{As} \approx k_{22} \cdot p_{tot} \cdot N^*_{As}\,, \tag{10b}$$

$$I_{dope} = \frac{v_{epi}}{F_1} \cdot (N_{As_{Si}})_{T_{epi}}\,, \tag{10c}$$

$$A^*_s \leftrightarrows As_{Si}\,, \qquad K_{sol} = \frac{(N_{As_{Si}})_{T_{epi}}}{N^*_{As}} \tag{10d}$$

The special incorporation step is characterized by a solution equilibrium (equ. (10d)).

In spite of dopant consumption by adsorption and layer doping, the equilibrium state within the gas is undisturbed, since the decomposed particles, adsorbed or incorporated, are reproduced by the feed of fresh dopant source substance in the gas stream. The mechanism derived above enables equ. (12/II) to be specified in another way given by equ. (11)

$$\frac{dN^*_{As}}{dx} = \frac{1}{v_{epi}} \cdot \left[k_{12} \cdot N_0 \cdot p^{eq}_{AsH_2} - k_{22} \cdot \left(p^{eq}_{H_2} + \frac{K_{sol}}{F_1 \cdot k_{22}} \cdot v_{epi}\right) \cdot N^*_{As}\right]. \tag{11}$$

From equ. (11) another hypothetical incorporation law is derived by integration

$$(N_{As^+_{Si}})_{tot,x} = \frac{K_{ad} \cdot K_{sol} \cdot N_0 \cdot p^0_{AsH_3}}{F_1 \cdot \left[p_{tot} + \frac{K_{sol}}{F_1 \cdot k_{22}} \cdot v_{epi}\right]} \times$$

$$\times \left[1 - \exp -\frac{k_{22}}{v_{epi}} \cdot \left(p_{tot} + \frac{K_{sol}}{F_1 \cdot k_{22}} \cdot v_{epi}\right) \cdot \Delta x\right]. \tag{12}$$

Not only the structural equality of the equs. (12) and (8), but also the impossibility of distinguishing them by evaluating doping experiments can be seen at a first glance. Doping experiments merely give rise to the quantification of the constants $A$, $B$, and

$L$ which are indiscriminately applicable to any of the hypothetical laws of dopant incorporation.

## 5. The consequent conception

Finally the hypothesis should be offered that dopant incorporation occurs as a single step reaction of the main component of the dopant source decomposition equilibrium. Arsenic, for example, requires the following reaction scheme:

$$AsH_2 \rightarrow As_{Si} + H_2\,, \qquad I_{13} = k_{13} \cdot p^{eq}_{AsH_2} \approx k_{13} \cdot p^{0}_{AsH_3}\,, \tag{13a}$$

$$H_2 + As_{Si} \rightarrow AsH_2\,, \qquad I_{23} = k_{23} \cdot p^{eq}_{H_2} \cdot (N_{As_{Si}})_{T_{epi}} \approx k_{23} \cdot p_{tot} \cdot (N_{As_{Si}})_{T_{epi}}\,, \tag{13b}$$

$$I_{dope} = \frac{v_{epi}}{F_1} \cdot (N_{As_{Si}})_{T_{epi}}\,, \tag{13c}$$

$$As_{Si} \rightarrow As^+_{Si} + e\,, \qquad K_{ion} = K^*_{ion} \cdot \frac{(N_{As^+_{Si}})_{T_{epi}}}{(N_{As_{Si}})_{T_{epi}}} = K^*_{ion} \cdot \frac{(1 - F_1)}{F_1}\,. \tag{13d}$$

In consequence of the above reaction scheme an additional hypothetical incorporation law can be formulated:

$$(N_{As^+_{Si}})_{tot,x} = \frac{K_{react} \cdot p^0_{AsH_3}}{F_1 \cdot \left[p_{tot} + \dfrac{v_{epi}}{F_1 \cdot k_{23}}\right]} \cdot \left[1 - \exp - \frac{k_{23}}{v_{epi}} \cdot \left(p_{tot} + \frac{v_{epi}}{F_1 \cdot k_{23}}\right)\right]. \tag{14}$$

It is not unexpected a result that the mathematical structure of equ. (14) is equal to that of the hypothetical laws already derived above. Additional equations may be derived applying the same pattern.

All of these hypothetical laws are true in a certain sense. They can be converted into the state of applicability to practice by introducing the empirical constants $A$, $B$, and $L$.

## 6. Nature and testability of the hypothetical dopant incorporation laws

After having substituted $A$, $B$, and $L$ for the theoretical constants of the various hypothetical laws all those laws change into the one empirically based, but theory-aided model equation similar to equ. (9/I). This is due to the general incorporation scheme the differential equation (12/II) is based upon: The dopant incorporation flux is balanced by two reaction steps, a forward directed reaction and its opposite directed one.

This general scheme is inherent to each of the incorporation mechanisms discussed above. It also belongs to the empirical equ. (9/I), for equ. (9/I) must be accepted as the solution of a differential equation belonging to the same type as equ. (12/II). In consequence the correct general incorporation mechanism was selected, though completely unaware of doing it, simultaneously when the empirical model equation had been formulated.

Finally let us consider the testability of the hypothetical laws. In case the numerical values of the empirical constants $A$, $B$, and $L$ could be reproduced from such a set of theoretical constants, the corresponding reaction scheme would prove to be true. Yet, testing is possible in a simpler way, for instance in the case of equ. (14). Figure 1 shows

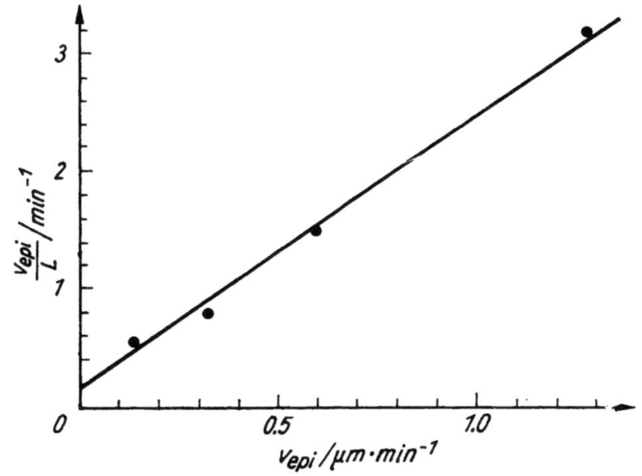


Fig. 1. Dependence of the empirical factor $L$ on layer growth rate $v_{epi}$. Data obtained by REIF et al. and evaluated by WONG et al.

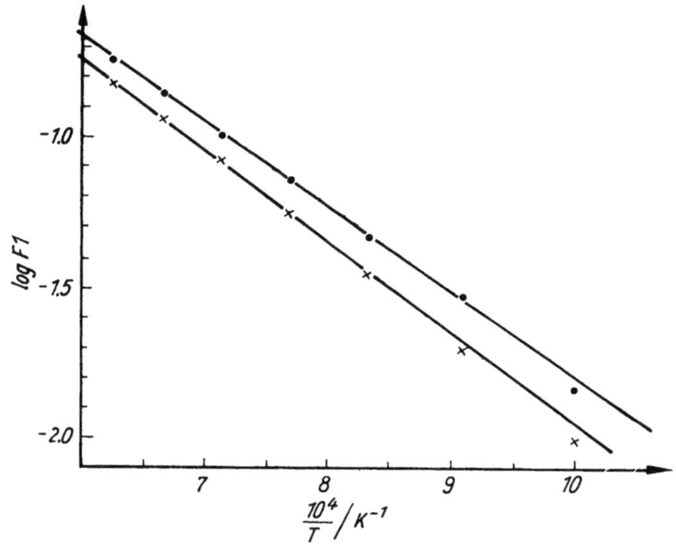


Fig. 2. Dependence of the term $F_1 = (N_{dope})_{T_{epi}}/(N_{dope})_{tot}$ on temperature, ······ $(N_{dope})_{to} = 10^{14}$ cm$^{-3}$, *** $(N_{dope})_{tot} = 10^{18}$ cm$^{-3}$

an experimental example of $L$ versus $v_{epi}$ leading to $F_1 = 0.44$ (c.f. equ. (4)), a value within the limits of the $F_1$-definition ($0 < F_1 < 1$). A stronger test requires the exact value of $F_1$. Figure 2 shows the result of a corresponding calculation. It demonstrates $F_1 \approx 0.1$ within the range of $1300 < T_{epi} < 1400$ K. Such a result, once accepted to be reliable, falsifies equ. (14) and together with it the underlying dopant incorporation mechanism. This result is also valid to equ. (3).

In case that a further test is required, our attention must then be focussed on the equilibrium and reaction rate constants. Equilibrium constants may be calculated from the basis data of thermodynamics applying equ. (15).

$$-R \cdot T \cdot \ln K = \Delta H - T \cdot \Delta S \qquad (15)$$

where: $\Delta H \equiv$ enthalpy of reaction, $\Delta S \equiv$ entropy of reaction.

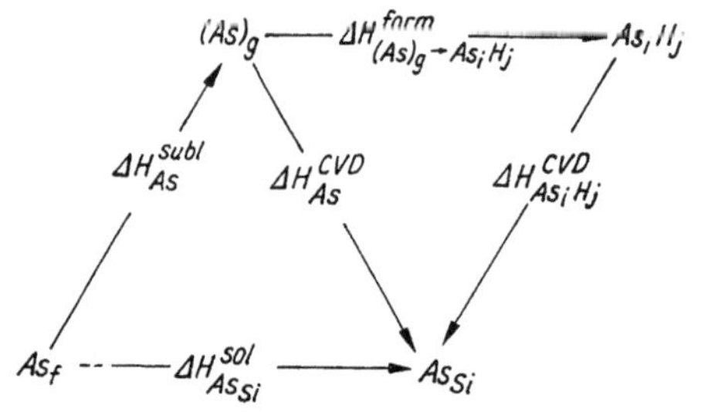


$$\Delta H^{CVD}_{As_iH_j} = \Delta H^{CVD}_{As} - \Delta H^{form}_{(As)_g \to As_iH_j} = \Delta H^{sol}_{As} - \Delta H^{form}_{(As)_g \to As_iH_j} - \Delta H^{subl}_{As}$$

Fig. 3. Enthalpy diagram demonstrating the connection of the incorporation enthalpies of $As_iH_j$ to the irrespective formation, sublimation and solution enthalpies of arsenic

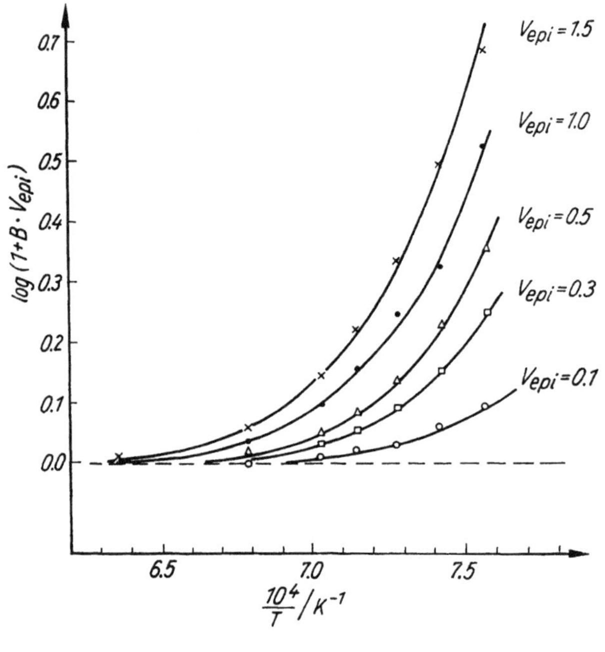


Fig. 4. Dependence of the to phosphorus incorporation related factor $B$ on temperature, the influence of layer growth rate taken into account as a parameter

Equ. (15) can be simplified by its differentiation with respect to the reciprocal temperature, since the enthalpy of reaction is then given by the temperature dependence of the equilibrium constant. The enthalpy of a special reaction is not only accessible from the reaction sequence considered, but also from roundabout reaction sequences. Thus the dopant incorporation enthalpy of a certain kind of particles, containing dopants, can be obtained irrespective of the incorporation reaction in case that the enthalpy of solution ($\Delta H^{sol}$), sublimation ($\Delta H^{subl}$), and formation ($\Delta H^{form}$) are known (KÜHNE). In Figure 3 arsenic sets an example how the incorporation enthalpies of different kinds of particles $As_iH_j$ differs from each other. Comparing experimental values with theoretical ones allows to select the special kind of particles which governs the overall incorporation reaction (KÜHNE et al.). In this way the incorporation mechanism expressed by equ. (3) can be distinguished from that characterized by equ. (14). But, as illustrated in Figure 3 equ. (8), for example, cannot be distinguished from equ. (14) by applying this method.

## 7. Conclusions

Theoretical foundations applied to epitaxial silicon layer doping enables us to select a general basis mechanism characterized by a forward and a backward directed reaction step both balanced by the dopant incorporation flux. Though true in principle, this general mechanism does not image incorporation in detail. It allows to derive a variety of hypothetical incorporation laws, which are not distinguishable by doping experiments, but are all transformable into one and the same model equation given, for instance, by equ. (8) with the empirical constants $A$, $B$, and $L$ substituted. That equation posses more explanatory force than the only empirically based equ. (9/I) due to its higher degree of theoretical content. The latter is demonstrated by using so called effective theoretical constants as demonstrated in equ. (16).

$$(N_{\mathrm{AsSi}})_{\mathrm{tot},x} = \frac{K^* \cdot p^0_{\mathrm{AsH_3}}}{(p_{\mathrm{tot}} + K^{**} \cdot v_{\mathrm{epi}})} \cdot \left[1 - \exp - \frac{k^*}{v_{\mathrm{epi}}} \cdot (p_{\mathrm{tot}} + K^{**} \cdot v_{\mathrm{epi}}) \cdot \Delta x\right]. \tag{16}$$

Because of its theoretical nature $K^{**}$ should rapidly decrease as the temperature is increased. Therefore $K^{**} \cdot v_{\mathrm{epi}}$ should be much smaller than the value of the total pressure, which simplifies equ. (16) in the range of higher deposition temperatures as given with equ. (16a). The predicted behaviour of $K^{**} \equiv B$, as shown in Figure 4, has been

$$(N_{\mathrm{AsSi}})_{\mathrm{tot},x} = \frac{K^*}{p_{\mathrm{tot}}} \cdot p^0_{\mathrm{AsH_3}} \cdot \left[1 - \exp - \frac{k^*}{v_{\mathrm{epi}} \cdot p_{\mathrm{tot}}} \cdot \Delta x\right] = K^* \cdot p^0_{\mathrm{AsH_3}} \tag{16a}$$

experimentally verified (KÜHNE). The right-hand term of equ. (16a) explains the empirical control equation (3/I) as that specific form of dopant incorporation valid in the range of higher deposition temperatures at atmospheric pressure under steady state conditions.

At low deposition temperatures the contradictory limiting case $K^{**} \cdot v_{\mathrm{epi}} \gg p_{\mathrm{tot}}$ should be correct. Then, restricted to steady state conditions, equ. (16) is simplified to:

$$(N_{\mathrm{AsSi}})_{\mathrm{tot}} = \frac{K^* \cdot p^0_{\mathrm{AsH_3}}}{K^{**} \cdot v_{\mathrm{epi}}} = \frac{k^*}{v_{\mathrm{epi}}} \cdot p^0_{\mathrm{AsH_3}} \,. \tag{16b}$$

In this way the theoretical content of equ. (16) exposes that the equilibrium-near process (c.f. 16a) and the irreversible chemical-reaction-near process (c.f. 16b) are not those incorrect ideas of dopant incorporation as formerly discussed in Part II of the paper, but must be considered to be true limiting cases of reality.

The original problem we started from concerned the question, how the dopant content in the epitaxial layer be controlled by the influence of the main parameters of the doping process. Equ. (16) represents the present state answere to that question. It describes the dopant incorporation process comprehensively as well as it explains the general mechanism of dopant incorporation. A full answere has been given with regard to the aspect of technology.

With regard to the aspect of science the problem, as yet, has not been solved completely, for the precise dopant incorporation mechanism has not been exposed. Further investigations to this end would lead us away from the practice-related aspect of the original problem. Thus we might let the unsolved part of the problem at rest until another practice-related problem arises which then is expected to be solved by applying to it the completed theory of dopant incorporation.

## References

BLOEM, J., GILING, L. J., GRAEF, M. W. M.: J. Electrochem. Soc. Amer. **121**, 1354 (1974)
GRAEF, M. W. M., LEUNISSEN, B. J. H., de MOOR, H. H. C.: ibid. **132**, 1942 (1985)
HERRING, R. B.: Electrochem. Soc. Fall Meeting, Low Angeles 1979
HINTZEN, H. T. J. M., BLOEM, J., GILING, L. J.: J. Electrochem. Soc. Amer. **131**, 1900 (1984)
KÜHNE, H.: Cryst. Res. Technol. **20**, 741 (1985)
KÜHNE, H.: ibid. **17**, 181 (1982)
KÜHNE, H.: ibid. **17**, 1217, 1379 (1982)
KÜHNE, H., SPERLING, R.: ibid. **15**, 403 (1980)
REIF, R., KAMINS, T. I., SARASWAI, K. C.: J. Electrochem. Soc. Amer. **126**, 644 (1979)
SHEPERD, W. H.: ibid. **115**, 652 (1968)
WONG, M., REIF, R.: IEEE Transact. on Electr. Devices **ED-32**, 83 (1985)
WONG, M., REIF, R., SRINIVASAN, G. R.: ibid. **ED-32**, 89 (1985)

(Received September 1, 1986)

*Author's address:*

Dr. H. KÜHNE
Institut für Halbleiterphysik der AdW der DDR
W.-Korsingstraße 2, PSF 409
Frankfurt/O.
DDR-1200

Cryst. Res. Technol. | 22 | 1987 6 | 777–781

Ch. Georgi

Ilmenau, DDR

# On the Origin of the (1$\bar{2}$16)-Orientation in Epitaxial AlN Layers on R-Plane Sapphire[1])

Epitaxial AlN grown by MOCVD on (01$\bar{1}$2)$Al_2O_3$ substrates usually is (1$\bar{2}$10)-oriented. Here for the first time (1$\bar{2}$16)AlN was found in MOCVD layers. This AlN orientation is known from other growth techniques, its origin is generally not understood. An explanation is given based on twinning in the early growth stages which is also applicable to the growth of (1$\bar{2}$16)ZnO on (01$\bar{1}$2)$Al_2O_3$.

Epitaktische AlN-Schichten, die mittels MOCVD auf (01$\bar{1}$2)$Al_2O_3$-Substraten abgeschieden wurden, sind gewöhnlich (1$\bar{2}$10)-orientiert. Hier wird erstmals das Auftreten von (1$\bar{2}$16)AlN in solchen Schichten beschrieben. Diese AlN-Orientierung ist von anderen Schichtablagerungsverfahren her bekannt, ihre Entstehung jedoch ungeklärt. Es wird eine Interpretation gegeben, die auf der Zwillingsbildung im Anfangswachstum der Schicht beruht und auch auf (1$\bar{2}$16)-orientiertes ZnO auf (01$\bar{1}$2)$Al_2O_3$ anwendbar ist.

## 1. Introduction

Aluminium nitride (AlN) is a wide-bandgap material ($E_g = 6.2$ eV; Yamashita et al.) which crystallizes in the wurtzite structure type. It is piezoelectric with a high acoustic velocity and is stable at high temperatures. Thus, heteroepitaxial AlN films grown by CVD techniques have a good potential for high-frequency surface acoustic wave devices (Duffy). The most useful substrate for such films is (01$\bar{1}$2)$Al_2O_3$ (also called R-plane sapphire) which yields the following epitaxial relationship (Manasevit et al.; Duffy et al.; Callaghan et al.; Liu et al.; Tsubouchi et al.):

$$(1\bar{2}10)\text{AlN} \parallel (01\bar{1}2)\text{Al}_2\text{O}_3\,, \qquad [0001]\text{AlN} \parallel [0\bar{1}11]\text{Al}_2\text{O}_3$$

AlN films grown by metal organic chemical vapor deposition (MOCVD) are reported to be exclusively (1$\bar{2}$10)-oriented (Duffy). In contrast, the additional orientation

$$(1\bar{2}16)\text{AlN} \parallel (01\bar{1}2)\text{Al}_2\text{O}_3$$

may be produced by the chloride process (Callaghan et al.; Kudakov et al.; Barovski et al.; Kiseleva et al.; Sokolov et al.) and by reactive molecular beam epitaxy (Yoshida et al.). Two equivalent azimutal orientations are possible:

$$[10\bar{1}0]\text{AlN} \parallel [0\bar{1}11]\text{Al}_2\text{O}_3 \quad \text{and} \quad [\bar{1}010]\text{AlN} \parallel [0\bar{1}11]\text{Al}_2\text{O}_3$$

(Callaghan et al.; Sokolov et al.). However, it has to be noted that this relation usually is not exactly realized, rather planes between (1$\bar{2}$16) and (1$\bar{2}$18) are parallel to the substrate surface, therefore it is sometimes denoted as (1$\bar{2}$17)-orientation (Yim et al.). A detailed investigation of the dependence of layer misorientation and azimutal

[1]) This study was carried out at TH Ilmenau, Department of Physics and Technology of Electronic Devices.

orientation on the substrate misorientation was carried out by SOKOLOV et al. for $(1\bar{2}16)$AlN on $(01\bar{1}2)Al_2O_3$.

Zinc oxide (ZnO), which is isotypic with AlN, also can be grown both in $(1\bar{2}10)$ and $(1\bar{2}16)$-orientation on R-plane sapphire by the chloride process (KASUGA, ISHIHARA; KASUGA, MOCHIZUKI; KASUGA et al.; note the obvious erroneous notation of the azimutal orientation in this papers, only the stereogram shown by KASUGA et al. exhibits the right relationship). LAU et al. obtained epitaxial ZnO on $(01\bar{1}2)Al_2O_3$ by MOCVD, from their unindexed RHEED-pattern follows $(1\bar{2}16)$ layer orientation too.

Reproducible relations between growth parameters and layer orientation could not be established till now for both materials. Only KUDAKOV et al. and KASUGA et al. related $(1\bar{2}16)$ to low, and $(1\bar{2}10)$ to high substrate temperatures. So the reasons for this ambiguous epitaxial behaviour of AlN and ZnO on $(01\bar{1}2)Al_2O_3$ still are not clear.

In this paper for the first time the occurrence of $(1\bar{2}16)$-oriented AlN in MOCVD films is reported. An explanation of the formation of $(1\bar{2}16)$-oriented AlN and ZnO on R-plane sapphire is given.

## 2. Experimental

Epitaxial AlN films were grown by the reaction of trimethylaluminium (TMA) and ammonia in the presence of hydrogen in a MOCVD apparatus established by EICHHORN and RENSCH. R-plane sapphire substrates of two different qualities (Tab. 1) were used in each run. The growth temperature was varied from 1373 to 1623 K, the TMA flow rate from 0.5 to $7.9 \cdot 10^{-5}$ mol $\cdot$ min$^{-1}$. Hydrogen and ammonia flow rates were maintained at 4750 and 970 cm$^3$ $\cdot$ min$^{-1}$, respectively. The orientation relationships were determined by reflection high energy electron diffraction (RHEED).

Table 1

Properties of R-plane sapphire substrates used in this work and the frequency of observed AlN layer orientations

| | | |
|---|---|---|
| quality | R1 | R2 |
| growth method | Czochralski | flame fusion |
| surface finish | chemomechanical | mechanical |
| range of substrate misorientation | $\pm 2°$ | $\pm 3°$ |
| dislocation density (cm$^{-2}$) | $5 \cdot 10^4$ | $4 \cdot 10^5$ |
| observed surface features | featureless | small angle grain boundaries, polishing scratches |
| frequency of observed AlN layer orientations (%): | | |
| $(1\bar{2}10)$ | 80 | 50 |
| $(1\bar{2}10) + (1\bar{2}16)$ | 18 | 34 |
| $(1\bar{2}16)$ | 2 | 16 |

## 3. Results and discussion

Epitaxial AlN films grown under the conditions given above showed both $(1\bar{2}10)$ and $(1\bar{2}16)$ orientation, often in the same layer (Fig. 1). No relations between observed orientation and growth parameters could be established. However, a remarkable connection between substrate quality and orientation distribution was found (Tab. 1): substrates of the poor quality R2 favour the growth of $(1\bar{2}16)$AlN.

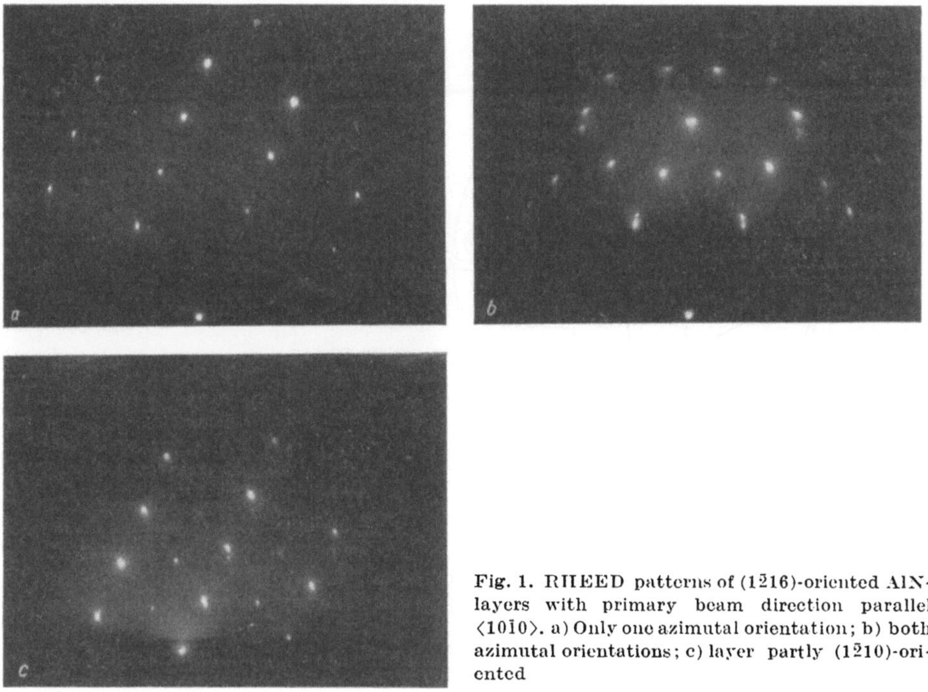

Fig. 1. RHEED patterns of $(1\bar{2}16)$-oriented AlN-layers with primary beam direction parallel $\langle 10\bar{1}0 \rangle$. a) Only one azimutal orientation; b) both azimutal orientations; c) layer partly $(1\bar{2}10)$-oriented

The orientation $(1\bar{2}16)$AlN $\|$ $(01\bar{1}2)Al_2O_3$ can not be described in terms of structural geometry (intergrowth of low-indexed planes of equal loading density, common structural elements of substrate and deposit, compatibility of symmetry elements) as it has been done for $(1\bar{2}10)$AlN on the same substrate by CALLAGHAN et al. This is evident also for possible "hidden" epitaxial intergrowths not parallel to the macroscopic interface like $(1\bar{1}01)$AlN $\|$ $(11\bar{2}0)Al_2O_3$ and (0001)AlN $\|$ $(\bar{1}2\bar{1}3)Al_2O_3$. Therefore it appears reasonable to regard the $(1\bar{2}16)$AlN orientation not as a true heteroepitaxial relationship but rather as a simulated one in the following way: First, heteroepitaxial nucleation of $(1\bar{2}10)$AlN takes place at the substrate surface. In the early growth stages of these nuclei twinning brings about an orientation change in such a way that the $(1\bar{2}16)$ plane of AlN is now parallel to the $(01\bar{1}2)$ substrate plane.

Nucleation and early growth will be strongly influenced by substrate surface defects (as they are present here above all in R2 substrates, see Tab. 1) and by the increased supersaturation which is possible before the CVD reactor reaches its stationary regime of diffusion controlled growth. On the other hand, growth twinning is known to be favoured in the nuclear stage, especially under conditions of supersaturation. So, for example, at steps parallel to $[0\bar{1}11]Al_2O_3$ created by substrate misorientation $(1\bar{2}10)$-oriented AlN nucleates under supersaturation $\Delta P_1$. The nulcei will grow preferentially along this steps in [0001]AlN main growth direction until they coalesce. Then further growth is blocked in this direction and supersaturation rises to a value $\Delta P_2 > \Delta P_1$, which allows nucleation between the steps too. For the nuclei already present at the steps this increasing supersaturation leads to an increasing probability of twin formation.

An analysis of the known twin laws for AlN shows that the $(1\bar{1}0\bar{2})$ twin described by GINDT and KERN is suitable to explain the change in orientation from $(1\bar{2}10)$ to $(1\bar{2}16)$ in the way proposed above (Fig. 2). From the observation of the same twin law for

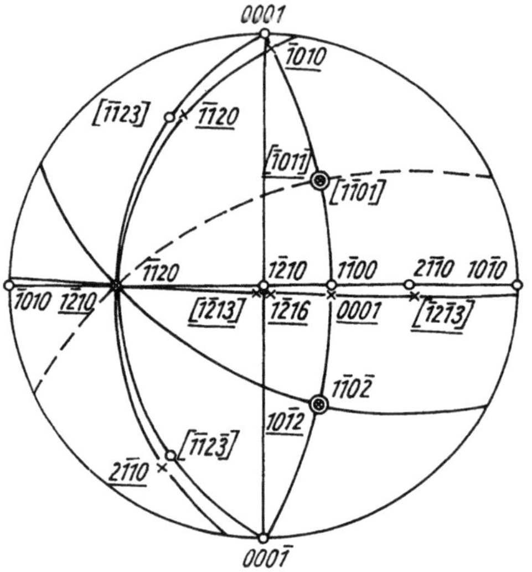


Fig. 2. Stereographic projection of the $(1\bar{1}0\bar{2})$ AlN twin. Projection plane is $(1\bar{2}10)$ of the first individual, indices of the other individual are underlined. Poles of twin elements are marked with additional circles, the trace of the twin plane is shown by a dashed line

ZnO (CESARO) and the almost perfect correspondence of the axial ratios of AlN ($c/a = 1.601$; SCHULZ and THIEMANN) and ZnO ($c/a = 1.600$; ABRAHAMS, BERNSTEIN) follows an analogous interpretation for $(1\bar{2}16)$-oriented ZnO.

The interpretation of the formation of $(1\bar{2}16)$-oriented AlN developed here is consistent with the dependence of azimutal orientation on substrate misorientation investigated by SOKOLOV et al. and with the relation between growth temperature and observed orientation reported by KUDAKOV et al. and KASUGA et al.: Starting from diffusion-controlled growth regime, the decreasing temperature leads to reaction-controlled growth, i.e. to a higher surface supersaturation, which favours the formation of $(1\bar{2}16)$AlN as discussed above.

## 4. Conclusion

Epitaxial AlN films of two different layer orientations $(1\bar{2}10)$ and $(1\bar{2}16)$ could be grown on $(01\bar{1}2)Al_2O_3$ by MOCVD, indicating that $(1\bar{2}16)$ AlN is not typical for any particular growth method. However, no relations between growth parameters and observed AlN orientation were found, only an influence of substrate quality was noted.

The $(1\bar{2}16)$ orientation cannot be interpreted as a true heteroepitaxial intergrowth, but it was shown here that it can be explained as the result of twin formation under the complex growth conditions given by the substrate surface and supersaturation in the eraly growth stages. This explanation can also be applied to $(1\bar{2}16)$-oriented ZnO on R-plane sapphire.

The author thanks his former colleagues from TH Ilmenau, Department of Physics and Technology of Electronic Devices, for supporting his work.

### References

ABRAHAMS, S. C., BERNSTEIN, J. C.: Acta cryst. **B25**, 2254 (1969)

BAROVSKI, N. V., KUDAKOV, U. D., SOKOLOV, E. B., SREDIN, U. G.: Izv. Akad. Nauk SSSR, Neorg. Mater. **19**, 1589 (1983)

CALLAGHAN, M. P., PATTERSON, E., RICHARDS, B. P., WALLACE, C. A.: J. Cryst. Growth **22**, 85 (1974)
CESARO, P.: Z. Krist. **24**, 618 (1895)
DUFFY, M. T.: in Heteroepitaxial Semiconductors for Electronic Devices, pp. 150–181, ed. by G. W. CULLEN and C. C. WANG, New York 1978
DUFFY, M. T., WANG, C. C., O'CLOCK Jr., G. D., McFARLANE III, S. H., ZANZUCCHI, P. J.: J. Electron. Mater. **2**, 359 (1973)
EICHHORN, G., RENSCH, U.: phys. stat. sol. **A 69**, K3 (1982)
GINDT, R., KERN, R.: Bull. Soc. Franc. Minéral. Cristallogr. **81**, 266 (1958)
KASUGA, M., ISHIHARA, S.: Jap. J. Appl. Physics **15**, 1835 (1976)
KASUGA, M., MOCHIZUKI, M., KOBAYASHI, K., SHIMIZU, A.: Jap. J. Appl. Physics **19**, 2289 (1980)
KASUGA, M., MOCHIZUKI, M.: J. Cryst. Growth **54**, 183 (1981)
KISELEVA, E. N., KUDAKOV, U. D., NAJDA, T. A., SOKOLOV, E. B., PASCIN, N. S., JAKOVKIN, I. B.: Mikroelektronika **12**, 481 (1983)
KUDAKOV, U. D., MALYUKOV, B. A., SOKOLOV, B. A., NAJDA, G. A.: in: Tez. I. Vsesojuz. konf. po fizike i techn. tonkich plenok, p. 92, Ivano-Frankovsk 1981
LAU, C. K., TIKU, S. K., LAKIN, K. M.: J. Electrochem. Soc. **127**, 1843 (1980)
LIU, J. K., LAKIN, K. M., WANG, K. L.: J. Appl. Physics **46**, 3703 (1975)
MANASEVIT, H. M., ERDMANN, F. M., SIMPSON, W.: J. Electrochem. Soc. **118**, 1864 (1971)
SCHULZ, H., THIEMANN, K. H.: Solid State Commun. **23**, 815 (1977)
SOKOLOV, E. B., MALYUKOV, B. A., KUDAKOV, U. D., NAJDA, G. A.: Cryst. Res. Technol. **18**, 53 (1983)
TSUBOUCHI, K., SUGAI, K., MIKOSHIBA, N.: 1982 Ultrasonics Symp. Proc., Vol. 1, p. 340
YAMASHITA, H., FUKUI, K., MISAWA, S., YOSHIDA, S.: J. Appl. Physics **50**, 896 (1979)
YIM., W. M., STOFKO, E. J., ZANZUCCHI, P. J., PANKOVE, J., ETTENBERG, M., GILBERT, S. L.: J. Appl. Physics **44**, 292 (1973)
YOSHIDA, S., MISAWA, S., FUJII, Y., TAKADA, S., HAYAKAWA, H., GONDA, S., ITOH, A.: J. Vac. Sci. Technol. **16**, 990 (1979).

(Received October 29, 1986)

*Author's address:*

Dr. Ch. GEORGI
Leninring 20
Ilmenau
DDR-6325

Cryst. Res. Technol. | 22 | 1987 | 6 | 782

**Book Review**

U. GONSER

# Microscopic Methods in Metals

Topics in Current Physics

Springer-Verlag, Berlin-Heidelberg-New York-Tokyo, 1986, 457 Seiten, 253 Abbildungen, ISBN 3-540-16251-8, DM 115.

Das vorliegende Buch, herausgegeben von U. GONSER, ist eine Zusammenstellung von 13 Berichten zur modernen Meßtechnik bzw. Festkörperanalytik. Darin wurden die metallkundlichen und metallphysikalischen Aspekte besonders herausgearbeitet und mit Beispielen belegt. Es ist U. GONSER gelungen, entsprechende Spezialisten und Mitautoren zu gewinnen, die es verstanden haben, auch komplizierte Zusammenhänge einfach darzustellen. So wurden theoretische und mathematische Abhandlungen auf ein Mindestmaß beschränkt und jeweils auf vertiefende Reviewarbeiten verwiesen. Natürlich gibt es wenig Bezüge der einzelnen Berichte aufeinander, so daß zwar jeder Bericht für sich gut lesbar ist, aber ein Lehrbuch im üblichen Sinne ist es nicht.

In den einzelnen Berichten werden aktuelle Themen hauptsächlichst unter atomistischen Aspekten abgehandelt und daraus die speziellen metallischen Eigenschaften interpretiert.

Im Einzelnen werden folgende Teilgebiete unter methodischer Sicht dargestellt: 1. Concerning Methods by U. GONSER; 2. Scanning Acoustic Microscopy by M. HOPPE and A. THAER; 3. High-Resolution Electron Microscopy by F. E. FUJITA and M. HIRABAYASHI; 4. Field Ion Microscopy by N. ERNST and G. EHRLICH; 5. X-Ray and Neutron Diffraction by C. N. J. WAGNER; 6. Extended X-Ray Absorption Fine Structure by D. G. STEARNS and M. B. STEARNS; 7. X-Ray Photoelectron Spectroscopy by G. K. WERTHEIM; 8. Auger Electron Spectroscopy by M. P. SEAH; 9. Positron Annihilation by W. TRIFTSHÄUSER; 10. Muon Spectroscopy by J. CHAPPERT and A. YAOUANC; 11. Perturbed Angular Correlation by TH. WICHERT and E. RECKNAGEL; 12. Nuclear Magnetic Resonance by P. PANISSOD; 13. Mössbauer Spectroscopy by U. GONSER.

Wie aus dieser Zusammenstellung ersichtlich, eine Palette z. T. modernster Verfahren und Meßtechniken, die sicher nicht in jedem Labor zugänglich sein werden. Darin zeigt sich aber auch der notwendige Trend zur Kooperation bei der Lösung metallphysikalischer Probleme, zudem eine Reihe von modernen Methoden (u. a. ELS, ESMA, SIMS) in dem vorliegenden Buch gar nicht abgehandelt werden konnten.

Das Buch vermittelt eine Fülle von Informationen über den Stand und die Trends der modernen Forschung unter metallphysikalischer Sicht. Es ist eine Hilfe für das Verständnis der unterschiedlichen Methoden und für die zu gewinnenden Aussagen. Es sollte zumindest in Bibliotheken der Hochschulen und Forschungszentren seinen Platz finden.

O. BRÜMMER

| Cryst. Res. Technol. | 22 | 1987 | 6 | 783–787 |
| --- | --- | --- | --- | --- |

N. S. Peev, Yu. M. Tairov, N. A. Smirnova, A. A. Kalnin

Institute of Microelectronic, Sofia, Bulgaria
Electrotechnical Institute, Leningrad, USSR

# Morphology of SiC Epitaxial Layers Grown by Temperature Gradient Zone Melting (I)

## Dependence of the Crystallization Boundary Smoothness on the Epitaxial Layers Growth Conditions

The dependence of the smoothness of the crystallization boundary on the temperature gradient zone melting conditions is investigated. The observed depth of the crystallization boundary is compared with the depth, calculated by the method proposed by D. E. Temkin.

В настоящей работе рассмотрена зависимость гладкости фронта кристаллизации от условия проведения процесса зонной плавки с градиентом температуры. Произведено сравнение экспериментально наблюдаемой глубины фазовой границы с вычисленной по методу Д. Е. Темкина.

### Introduction

The problems of the dependence of the epitaxial layers properties on their growth conditions are discussed in many papers (Chernov, Ljubov; Ugai; Lodis, Parker; Burton et al.; Kazumesa et al.). Because by knowing this dependence one may influence the epitaxial layers properties by changing their growth conditions. It is indicated as well (Peev et al.) that no such dependence of the electrophysical properties of the layers — carrier mobility, concentration and compensation degree on their growth conditions — process temperature, initial melt composition and the substrate side (Si- or C-side) was observed. Another series of experiments has shown, however, that the structural properties of the layers — their general structural perfection and the quantity of the second phase inclusions display such dependence on their growth conditions. During the experiments we observed a clearly expressed relationship between the crystallization boundary smoothness and the layers growth conditions. This very relationship is the object of discussion in the first paper. There is a need therefore to explain at least qualitatively this relationship.

The epitaxy technique is described in the work (Peev et al., Part II).

### Discussion

The precise definition of the crystallization boundary morphology during the liquid phase epitaxy is not possible because this boundary is continuously covered by the liquid phase layer. The observation of the crystallization boundary is possible only after cooling of the grown structures and elimination of the growth melt. This means that the last portions of the epitaxial layer are deposited in conditions far from typical for the given process. It is not clear therefore whether the observed crystallization boundary morphology is formed during the growth of the epitaxial layer or during the cooling of the structure.

In our experiments a relationship (with high degree of reproducibility) between the smoothness of the crystallization boundary and the process conditions was established. By taking into account that the thickness of the liquid phase was in the range of 30 to 50 μm, that the growth rate reached 0.1—1 μm/min and that the considered maximum roughness of the crystallization boundary reached 100 μm (several times larger than the zone thickness) a conclusion was made that such large deviation from the planarity of the phase boundary were impossible to emerge in such a short time as the heater took only 1—2 min for cooling. Therefore the observed crystallization boundary morphology characterized the epitaxial layers growth process itself.

The smoothness of the crystallization boundary is a very important feature of the growth process and it may serve as a measure of the quality of the apitaxial layers. During the investigations of the grown layers their C–V curves were measured and it was established that in some samples the quantity $N_D - N_A$ measured in different regions of the layer surface differed two or three times.

The problem of inhomogeneous layer doping parallel to the crystallization boundary is discussed in relation to its quality by CHERNOV and LYUBOV. This approach helps the qualitative understanding of process of epitaxial growth. The same problem (concerning smoothness of crystallization boundary) is discussed in many other works (JACSON; TEMKIN 1964 and 1965; BURTON et al.). It must be noted that BURTON-CABRERA-FRANK's approach is suitable for gas phase growth process, where the environment slightly influences the interface boundary morphology and in this case the surface morphology may be considered only as a function of the growth temperature.

Free from these deficiency are the approaches of JACSON and of TEMKIN. By the crystallization process activation energy $\Delta E$, which is accepted to be equal to the process of solution activation energy, the environment influences directly on the phase boundary perfection. Moreover, the parameters $w$ and $\varepsilon_{11}$, $\varepsilon_{12}$, $\varepsilon_{22}$ account the influence of the crystallographic structure and crystallographic direction on the crystallization boundary smoothness. By the TEMKIN approach the influence of the supersaturation on the crystallization boundary smoothness is taken into account, which makes it even

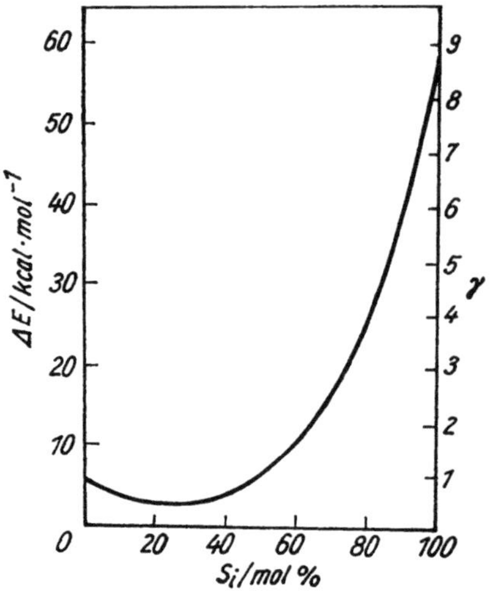


Fig. 1. Dependence of the parameters $\Delta E$ and $\gamma$ on initial melt composition

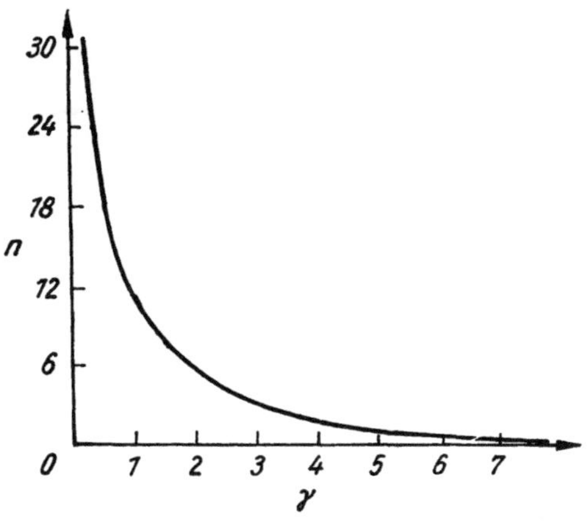


Fig. 2. Dependence of the interface boundary depth (given by the number $n$ of the monoatomic layers) on the parameter $\gamma$

more suitable for the epitaxial growth process. Another difference between these approaches is the criterion, which they use for the crystallization boundary smoothness definition. The ratio $N_A/N$, where $N_A$ is the number of the particles already deposited on the surface and $N$ is the number of all free places on the smooth crystal surface, is used as criterion for the interface smoothness in the JACSON's approach. If the value of this ratio is approximately 0.5 then the interface boundary is uneven but if this value is close to 0 or 1 then this boundary is smooth. In the TEMKIN approach another criterion is proposed: the depth of the interface boundary, expressed in the number of monoatomic layers — if this number is bigger, the interface boundary is uneven, and if it is smaller, then the boundary is smooth.

It must be noticed that both approaches consider the interface boundary smoothness on the atomic level. In our experiments it was impossible to determine the interface boundary smoothness on atomic level. Nevertheless, the practically observed macrodefects on the crystallization boundary are compatible with the TEMKIN approach.

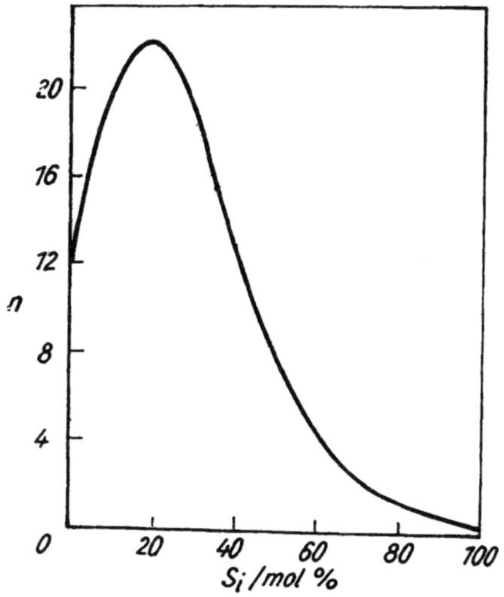


Fig. 3. Dependence of the interface boundary roughness $n$ on the initial melt composition

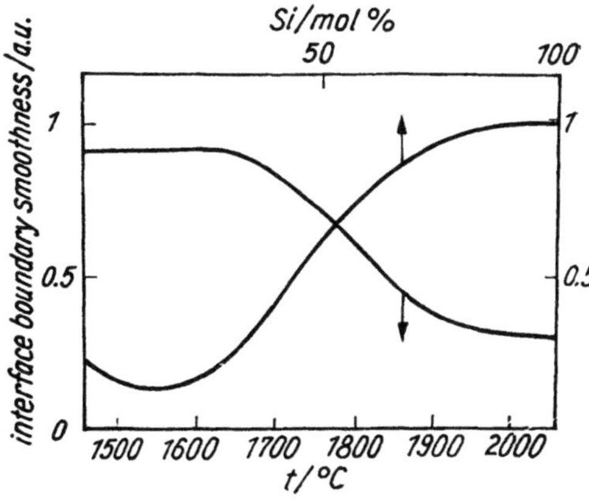


Fig. 4. Dependence of the interface boundary smoothness on the growth temperature and on the initial melt composition (experimental observed)

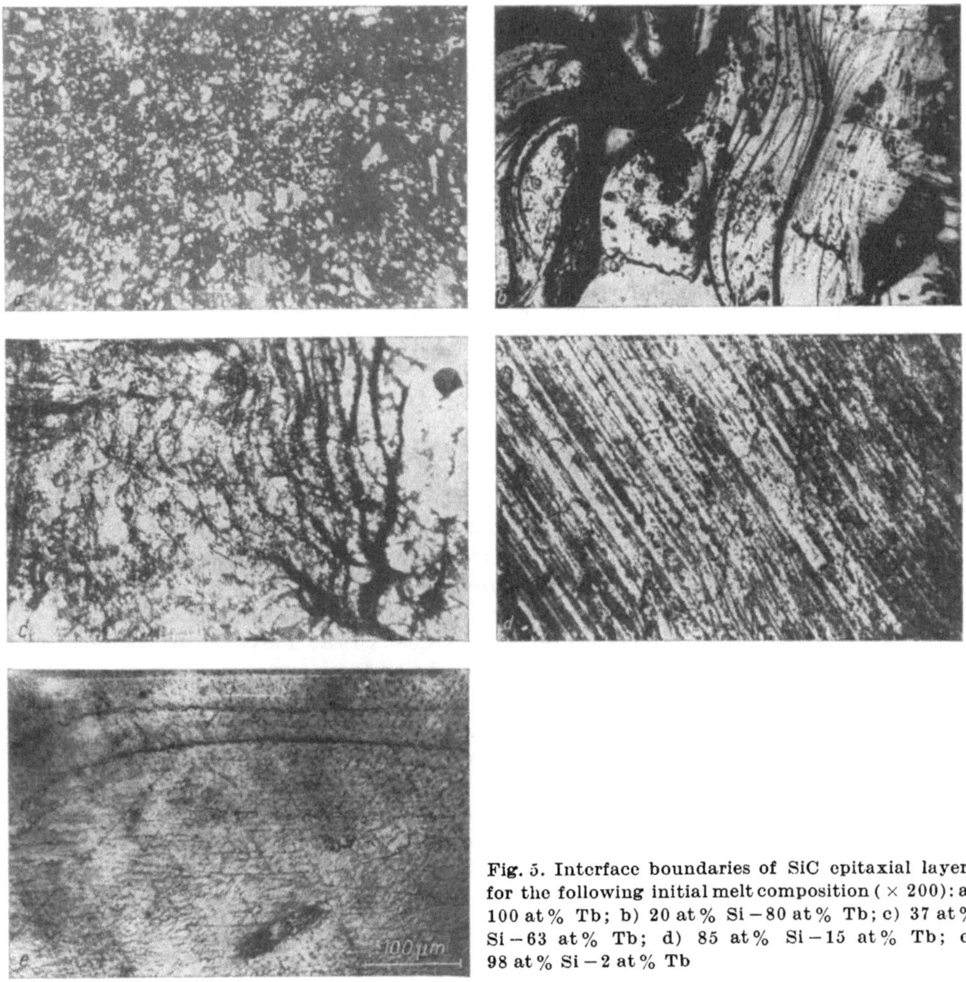


Fig. 5. Interface boundaries of SiC epitaxial layers for the following initial melt composition ( × 200): a) 100 at% Tb; b) 20 at% Si – 80 at% Tb; c) 37 at% Si – 63 at% Tb; d) 85 at% Si – 15 at% Tb; e) 98 at% Si – 2 at% Tb

Such a coincidence may be accidental, but it may be of a natural cause, therefore it deserves attention.

The TEMKIN approach will be more fulli discussed because it better coincides with the results from the experiments. The dependence of the interface boundary depth on the parameter $\gamma = 0.67 \cdot \Delta E/RT$ and the activation energy $\Delta E$ on the initial melt composition (Figs. 1 and 2) (TAIROV et al.) are used for the calculation of the depth of the crystallization boundary. The obtained results are given in Figure 3 and shown that in the Si–Tb–SiC system, beginning from the point 100 at% the crystallization boundary become more uneven and in the region around the composition 80 at.% Tb–20 at.% Si a maximum of the interface boundary depth is observed. This means that the crystallization boundary smoothness is worst. After further increase of the Si contents in the melt composition the smoothness of the interface boundary increases and at Si concentration above 70 at.% it reaches maximum — this means that the crystallization surface defects are commensurable with the interatomic distance.

From the expression $\gamma = 0.67\Delta E/RT$ and from the dependence of the interface boundary depth $n$ on the parameter $\gamma$ (Fig. 3, TEMKIN 1964 and 1965) follows, that the decrease of the growth temperature leads to the increase of the crystallization boundary smoothness. This conclusion which is based on the TEMKIN's approach, was confirmed in our experiments (Fig. 4).

During our experiments a reproducible dependence of the crystallization boundary smoothness on the initial melt composition and on the growth temperature was observed. In Figure 4 the interface smoothness is marked on the ordinate axis and the photos from Figure 5 as an illustration are shown. The comparison of the observed macrodefects, given in Figures 4 and 5, with the calculated interface boundary depth on atomic level, shown in Figure 3, displays a principal coincidence of the obtained results. This coincidence may be accidental, but it is quite possible, that they are interdependent each of another. If this proves to be true, the described approach will give the possibility to determine those liquid phase compositions, which will assure the best smoothness of the interface boundary of a given crystal.

## References


BURTON, W., CABRERA, N., FRANK, F.: Phil. Trans., **A243**, 299 (1951)

CHERNOV, A. A., LYUBOV, B. Ja.: in: Rost crystallov, Moscow, V. 5 (1965), p. 11

JACSON, K. A.: in: Shidkie metalli i ich zatverdevanie, 1962, p. 200

KAZUMEZA, H., KAZUJOSHI, T., NOBUHIKO, S., ISAMU. A.: Jap. J. Appl. Physics **23**, No. 1 68 (1984)

LODIS, R., PARKER, R.: Rost monocrystallov, Moscow 1974, p. 540

TATROV, Yo. M., PEEV, N. S., SMIRNOVA, N. A., KALNIN, A. A.: Crystal Res. Technol. **21**, 1503 (1986)

PEEV, N. S., TAIROV, Yu. M., SMIRNOVA, N. A., KALNIN, A. A.: Crystal Res. Technol. **21**, 1509 (1986)

PEEV, N. S., TAIROV, YU. M., SMIRNOVA, N. A., KALNIN, A. A.: Cryst. Res. Technol, **22**, 59 (198)

TEMKIN, D. E.: in: Rost crystallov, V. 5, Moscow 1965, p. 89

TEMKIN, D. E.: in: Mechanismi i kinetica crystallizatsii, Moscow 1964, p. 86

UGAY, Ya. A.: Vvedenie v chemie polupr., Moscow 1965, p. 314


(Received November 4, 1986)

*Authors' addresses*:

Dr. N. S. PEEV
Institute of Microelectronics
Boulvd "Lenin" 7 km
Sofia — 1784
Bulgaria

Prof. Dr. Yu. M. TAIROV, Dr. N. A. SMIRNOVA, Dr. A. A. KALNIN
Electrotechnical Institute, Dept of Dielectrics and Semiconductors
ul. "Prof. Popov", 5
Leningrad — 197022
USSR

**Book Review**

V. D. SCOTT, G. LOVE (Eds)

## Quantitative Electron-Probe Microanalysis

Ellis Horwood Ltd., Chichester (England), 1983, 131 figures, 31 tables, 345 p., approx. 10 £ (paperback), ISBN 0-85 312-672-0

The book stems from a series of Summer Schools held regularly at the University of Bath (Great Britain) to train scientists with differing backgrounds in this important method of materials characterization. Insofar one could expect the book to be an introduction into electron probe microanalysis. But the book is by far more a "Monograph" than an "Introduction". The authors M. G. C. Cox, G. LOVE, T. MULVEY, S. J. B. REED, and V. D. SCOTT are well-known experts in the field. Thus, it is no wonder that from such a cooperation a book results which represents an actual and comprehensive status report on X-ray microanalysis.

On the other hand, writing a book by a group of authors is — as a necessity — not free from some problems. Intersections and repetitions are unavoidable. For example, Fig. 9.1 is nearly identical with Fig. 2.19. And the peak-to-background technique is described in chapter 11.5 on page 260 f as well as in chapter 13.3 on p. 318 f (without reference to chapter 11.5). Furthermore, valuable data and references may not in any case be found in chapters where the reader is expecting them. As an example, the results of COATES regarding the relative intensities of M-lines are mentioned in chapter 5.7 (light-element analysis) instead of chapter 2.2 (X-ray emission spectra). Insofar, the book does not reach the same level with respect to arrangement and conciseness as REED's excellent monograph from 1975. But the special value of the new book is, to the reviewer's opinion, that it contains a very large amount of that what happened within these 8 years. In this context the numerous references are of great value.

I had special pleasure when reading chapter 1: "The Development of Electron-Probe Microanalysis — an Historical Perspective", written by TOM MULVEY. This is really a brillant and fair perspective with a lot of details which are hard to find in other textbooks on this subject.

In technical respect the book is well printed with clear figures and a minimum of misprints. The only point of criticism is that — due to the narrow binding — the inner parts of the pages are a little difficult to read.

Everyone who has a scientific interest in electron-probe microanalysis should possess this book in his personal library.

M. WENDT

| Cryst. Res. Technol. | 22 | 1987 | 6 | 789–792 |
|---|---|---|---|---|

K. Sangwal

Pedagogical University of Częstochowa, Poland

# A New Equation for the Temperature Dependence of Density of Saturated Aqueous Solutions of Electrolytes

A new empirical three-parameter equation $d = d_0 \exp[\beta'(T - T_0)^2]$ (where $\beta'$, $d_0$, and $T_0$ are constants for a salt) has been proposed to describe the dependence of density $d$ (g/ml³) of saturated aqueous solutions of electrolytes on temperature $T$ (K) in the whole range of temperature in which solubility increases or decreases with an increase in temperature.

In the temperature interval in which density $d$ of saturated solution of a salt increases or decreases, its temperature dependence is usually described by the empirical relations (Sokołowski; Karniewicz et al.; Szewczyk et al.)

$$d = d_{01} + \alpha_1 t + \beta_1 t^2 \,, \tag{1}$$

$$d = d_{02} + \alpha_2 T + \beta_2 T^2 \,, \tag{2}$$

where $d_{01}$ and $d_{02}$ are the densities of the solution at $t = 0$ and $T = 0$, respectively, and $\alpha_1$, $\alpha_2$, $\beta_1$, and $\beta_2$ are experimentally determined constants. Equation 2 may be transformed into Equation 1 because $d_{01} = d_{02} + 273\alpha_2 + (273)^2\beta_2 + 546\beta_2 t$. It is found (Sokolowski; Szewczyk et al.) that usually $\alpha_1 \gg \beta_1$ and $\alpha_2 \gg \beta_2$.

The temperature dependence of density of constant-concentration solutions is also usually represented by Equations 1 and 2. Moreover, one notes that when the quadratic terms in the above equations are neglected, they take a form which is similar to that of the coefficient of thermal expansion of liquids and solids. However, the temperature dependence of density of saturated solutions is relatively more complicated than that for constant-concentration solutions or for pure solvents because of the simultaneous variation of solute concentration with temperature. The quadratic terms also cannot be neglected when density data covering a large temperature interval are to be represented.

In this paper a new empirical three-parameter equation is proposed to describe the temperature dependence of density of saturated aqueous solutions of electrolytes.

Raw experimental data for the analysis of the temperature dependence of density of saturated solutions of electrolytes were taken from the literature (Postinikov; Mullin; Sokołowski; Szewczyk et al.). In all 35 salts were considered (Tab. 1). The data were analysed according to the equation

$$d = d_0 \exp[\beta'(T - T_0)^2] \,, \tag{3}$$

where $d_0$, $\beta'$, and $T_0$ are constants characteristic for an electrolyte and $T$ is the temperature in degrees Kelvin. After arbitrarily taking different values of $T_0$, its preliminary values for various salts were determined from the plots of $\ln d$ against $(T - T_0)^2$. The value of $T_0$ that yielded the best straight line fit for the plots of $\ln d$ against $(T - T_0)^2$ was selected. Using the selected values of $T_0$, the values of the parameters $d_0$ and $\beta'$ were calculated by using a computer programme in which $T_0$ was varied in

Tablo 1

Values of constants $T_0$, $d_0$, $\beta'$, $\beta$ and correlation coefficient $r$ for some salts

| Compound | data source | $T_0$/K | $d_0$/g/ml³ | $10^6\beta'/K^{-2}$ | $10^{11}\beta$ | $r$ |
|---|---|---|---|---|---|---|
| LiCl | a | 223 | 1.2832 | 2.1619 | 4.34 | 0.9949 |
| NaCl | a | 213 | 1.2142 | −1.6804 | − 3.70 | 0.9932 |
| $NaNO_3$ | a | 123 | 1.2731 | 2.8971 | 19.14 | 0.9966 |
| $NaNO_2$ | a | 123 | 1.2589 | 2.2994 | 15.19 | 0.9939 |
| NaOH | a | 273 | 1.5316 | 12.1566 | 16.31 | 0.9990 |
| $NaHCO_3$ | a | 203 | 1.0394 | 2.7807 | 6.74 | 0.9967 |
| $Na_2CO_3$ | a | 243 (<313) | 1.2103 | 23.0499 | 39.03 | 1 |
| | | (>313) | 1.3838 | −4.3783 | − 7.41 | 0.9988 |
| $Na_2SO_4$ | a | 193 (<313) | 0.6610 | 55.9040 | 150.08 | 1 |
| | | (>313) | 1.3913 | −3.5885 | − 9.63 | 0.9922 |
| | b | 263 (<313) | 1.0069 | 148.8291 | 215.16 | 0.9983 |
| | | 253 (>313) | 1.3542 | −7.3224 | −11.43 | 0.9969 |
| $Na_2SO_3$ | a | 213 (<313) | 1.1184 | 14.2322 | 31.36 | 1 |
| | | (>313) | 1.3459 | −5.4409 | −11.99 | 0.9997 |
| $Na_3PO_4$ | a | 243 | 0.9973 | 34.4507 | 58.34 | 0.9994 |
| KCl | c | 123 | 1.1305 | 1.2442 | 8.22 | 0.9549 |
| | a | 123 | 1.1306 | 1.3064 | 8.63 | 0.9830 |
| KI | a | 173 | 1.6408 | 3.0141 | 10.07 | 0.9703 |
| $KClO_4$ | c | 253 | 1.0033 | 4.5300 | 7.07 | 0.9983 |
| $KNO_3$ | a | 143 | 0.8862 | 12.2690 | 59.99 | 0.9944 |
| $KCH_3CO_2$ | c | 273 | 1.3984 | 9.1293 | 12.24 | 0.9989 |
| $K_2CO_3$ | c | 223 | 1.5373 | 2.2672 | 4.55 | 0.9851 |
| $K_2Cr_2O_7$ | d | 203 | 0.9473 | 15.4734 | 37.54 | 0.9830 |
| $K_3[Fe(CN)_6]$ | a | 133 | 1.1185 | 2.3438 | 13.25 | 0.9941 |
| $K_4[Fel(CN)_6]$ | a | 133 | 1.0432 | 4.3481 | 24.58 | 0.9853 |
| $NH_4C_2$ | c | 133 | 1.0516 | 0.8001 | 4.52 | 0.9012 |
| $NH_4C_2O_4H$ | c | 283 | 1.0168 | 11.0987 | 13.85 | 0.9873 |
| $(NH_4)_4SO_4$ | c | 233 | 1.2420 | 0.8628 | 1.58 | 0.9810 |
| RbClO | c | 263 | 1.0049 | 5.2947 | 7.65 | 0.9978 |
| $MgCl_2$ | a | 223 | 1.3265 | 1.3275 | 2.66 | 0.9975 |
| $CoSO_4$ | a | 93 | 1.0856 | 5.2590 | 60.80 | 0.9933 |
| $CuCl_2$ | a | 233 | 1.4925 | 5.0259 | 9.25 | 0.9997 |
| $CuSO_4$ | a | 243 | 1.1435 | 19.6440 | 33.26 | 0.9999 |
| $PbCl_2$ | a | 223 | 1.0156 | −1.3283 | −2.67 | 0.9909 |
| | c | 223 | 1.0128 | −1.0485 | −2.10 | 0.9917 |
| $PbBr_2$ | c | 283 | 1.0064 | −1.1888 | −1.48 | 0.9886 |
| $PbI_2$ | c | 253 | 1.0032 | −2.8409 | −4.43 | 0.9987 |
| $Pb(NO_3)_2$ | a | 123 | 1.1217 | 8.2179 | 54.31 | 0.9903 |
| $Al(NO_3)_3$ | a | 273 | 1.3739 | 20.3254 | 27.27 | 0.9994 |
| $KAl(SO_4)_2$ | a | 283 | 1.0431 | 54.2427 | 67.72 | 0.9988 |
| $NH_4Al(SO_4)_2$ | a | 283 | 1.0407 | 44.5820 | 55.66 | 0.9997 |
| $CsAl(SO_4)_2$ | c | 283 (<313) | 1.0019 | −2.1255 | −2.65 | 0.9284 |
| | | (>313) | 0.9641 | 12.6380 | 15.77 | 0.9587 |

Data source: (a) POSTNIKOV, (b) SOKOŁOWSKI, (c) MULLIN, (d) SZEWCZYK et al.

steps of 10 K to get the best correlation coefficient. The values of parameters $T_0$, $d_0$, and $\beta'$ for various electrolytes are given in Table 1.

It should be noted that, like the constants of Equations 1 and 2, the constants $T_0$, $d_0$, and $\beta'$ are unique for an electrolyte in the temperature interval in which solubility increases or decreases. The value of $d_0$ is related with the solubility of an electrolyte while that of $\beta'$ depends on the temperature coefficient of solubility (Tab. 1). In the case of $T_0$, on the other hand, no such connection with solubility is apparent. Moreover, the determination of the exact value of $T_0$ often remains a point of controversy be-

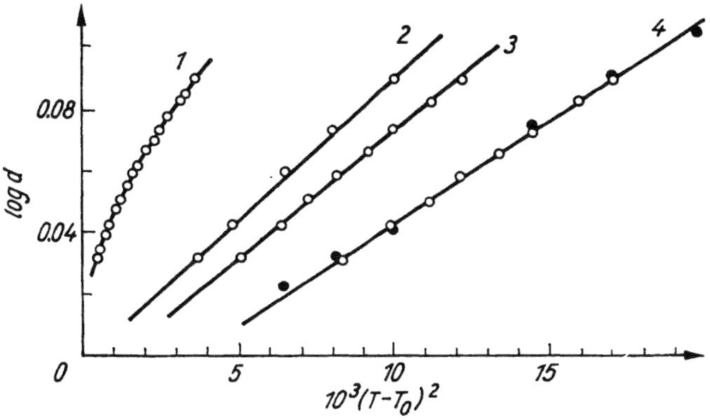


Fig. 1. Plots of log $d$ against $(T - T_0)^2$ for saturated solutions of $K_2Cr_2O_7$ with various values of $T_0$: (1) 273, (2) 233, (3) 223 and (4) 203 K. For the sake of clarity, only a small number of points are shown. The filled circles in curve 4 represent the data from POSTINIKOV, while all other data are taken from SZEWCZYK et al. to illustrate the point that data from different sources may give differing values of constants $d_0$, $T_0$, and $\beta'$

Table 2

Values of parameters $\beta'$, $\beta$ and $d_0$ corresponding to various selected values of $T_0$ for $K_2Cr_2O_7$ saturated aqueous solutions

| $T_0$/K | $d_0$/g/ml³ | $10/\beta'/K^{-2}$ | $10^{11}\beta$ |
|---|---|---|---|
| 273 (>323) | 1.0975 | 31.08 | 41.71 |
| (<303) | 1.0306 | 79.44 | 106.59 |
| 233 | 0.9953 | 21.48 | 39.57 |
| 223 | 0.9774 | 19.30 | 38.80 |
| 203 | 0.9473 | 15.47 | 37.54 |

cause it strongly depends on factors such as measurement errors and purity of solute and solvent. However, another dimensionless parameter $\beta = \beta'/T_0^2$ is still a measure of the behaviour of various electrolytes. This coefficient is relatively poorly dependent on the exact choice of $T_0$. For practical purposes the value of $\beta$ may be estimated for $T_0 = 273$ K from the density data in a small temperature range at elevated temperatures (Fig. 1). These features of $\beta$-coefficient may be noted from Table 2 for $K_2Cr_2O_7$.

Density data are frequently required for the determination of solute concentration and supersaturation during crystallization. Equation proposed in this work and the parameters $d_0$, $T_0$, $\beta'$, and $\beta$ given in Table 1 may be hoped to find their use for this purpose, particularly at high temperatures where it is difficult to determine density due to technical reasons.

This work was partially financed under Research Project CPBP 0.12.

## References

KARNIEWICZ, J., KOLASIŃSKI, W., SZEWCZYK, J.: J. Cryst. Growth **57**, 462 (1982)
MULLIN, J. W.: Crystallization, 2nd edition, London 1972
POSTINIKOV, V. A.: Zh. Fiz. Khimii **44**, 236 (1970)

SOKOŁOWSKI, T.: J. Cryst. Growth **52**, 829 (1981)
SZEWCZYK, J., SOKOŁOWSKI, W., SANGWAL, K.: J. Chem. Engng. Data **30**, 243 (1985)

(Received October 10, 1986)

*Authors' address:*

Dr. hab. K. SANGWAL
Department of Physical and Solid State Chemistry,
Pedagogical University of Częstochowa
Al. Zawadzkiego 13/15
42 200 Częstochowa
Poland

| Cryst. Res. Technol. | 22 | 1987 | 6 | 793–798 |
|---|---|---|---|---|

P. SEDLACEK, R.-J. KUBAN, K.-O. BACKHAUS

Zentralinstitut für physikalische Chemie der Akademie der Wissenschaften der DDR, Berlin

# Structure Determination of Polytypes (I)

Crystal structure determination of polytypes with diffraction methods can be divided into two steps. The first step is the structure determination of the building unit of the polytypic substance (in most cases layers). The second step is the determination of the stacking sequence of the corresponding building units. The stacking can be ordered as well as disordered. Various methods for solving these problems are described. The aim of this paper is to treate these methods from a uniform point of view.

Die Kristallstrukturbestimmung mit Beugungsmethoden von Polytypen kann in zwei Schritte eingeteilt werden. Der erste Schritt ist die Strukturbestimmung der Baueinheit der polytypen Substanz (in den meisten Fällen Schichten). Der zweite Schritt besteht in der Bestimmung der Stapelfolge der entsprechenden Baueinheiten. Die Stapelung kann sowohl geordnet als auch fehlgeordnet sein. Verschiedene Methoden zur Lösung dieser Probleme sind beschrieben worden. Das Ziel dieser Arbeit ist es, diese Methoden von einem einheitlichen Standpunkt zu behandeln und zu vergleichen.

## 1. Introduction

According to the definition recommended by the ad hoc-Nomenclature-Committee of the International Union of Crystallography and the International Mineralogical Association: "Polytypism is the phenomenon of the existence of an element or compound in two or more layer-like crystal structures that differ in layer stacking sequences. The layers need not be crystallographically identical, but should be similar . . ." (BAILEY et al. (1977 a, b).

For structures consisting of layers a general structure factor relation is given by

$$F_{(h,k,\zeta)} = \sum_{s} S^{(s)}_{(h,k,\zeta)} F^{(s)}_{(h,k,\zeta)}\,, \tag{I}$$

where $F^{(s)}_{(h,k,\zeta)}$ is the structure factor of the single layer of sort (s) and $S^{(s)}_{(h,k,\zeta)}$ is the function characterizing the stacking of the layers. If there is no special information, than the following quantities $I$, $F$, $K$, $S$ and $\varphi$ depend on $(h, k, \zeta)$, where $\zeta$ is a continuous variable in opposite to $h$ and $k$. All translationally equivalent layers may belong to one sort of layers (SEDLACEK 1978). Geometrically equivalent layers belong to one kind of layers (JAGODZINSKI 1949a). From (I) the intensity relation within coherent regions is given by

$$|F|^2 = \sum_{s} \sum_{t} \tilde{S}^{(s)} S^{(t)} \tilde{F}^{(s)} F^{(t)} \tag{II}$$

with

$$|S^{(s)}|^2 = S^{(s)} \tilde{S}^{(s)}\,,$$

$$|F^{(s)}|^2 = F^{(s)} \tilde{F}^{(s)}$$

and for the whole crystal as sum over all coherent regions by

$$I = \sum_{s} \sum_{t} K^{(s,t)} \tilde{F}^{(s)} F(t) \, . \tag{III}$$

One of the inherent properties of polytypic substances is the appearance of one-dimensional disorder. In the following ordered polytypes are regarded as special cases of disordered polytypes. Many specialists in this field have developed methods for characterization and calculation of the intensity distribution of onedimensionally disordered crystals (JAGODZINSKI 1949 a, b, c, 1954; WILSON 1942, 1943; HENDRICKS, TELLER 1942; ZACHARIASEN 1947; MERING 1949; PATTERSON 1952; KAKINOKI, KOMURA 1952 a, b, 1954, 1962, 1965; KAKINOKI 1967; GEVERS 1952, 1954 a, b; WARREN 1955, 1959; MEGAW 1960; JOHNSON 1963; ALLEGRA 1961, 1964; SATO 1966; LELE et al. 1969; LELE 1969; PRASAD, LELE 1971; TAKAKI et al. 1976; TAKAKI 1977; VAN DER PERRE et al. 1980; ONODA, KAWADA 1980; ONODA et al. 1980). These methods base on probability parameters only, neglecting energetical considerations. Therefore they are generally applicable, but require preliminary informations about the building principle of the structure. They are not so well suited for structures with unknown structural principle (structure of any kind of layer and of any kind of layer pair).

The Theory of OD-structures, developed by DORNBERGER-SCHIFF and GRELL-NIEMANN (1961), DORNBERGER-SCHIFF (1964, 1966, 1979, 1980) is motivated by energetical principles, reducing the number of stacking possibilities to energetically probable possibilities, explained by geometrical considerations. As far the most polytypic structures which have come to our knowledge, have turned to be OD-structures, the results of OD-theory may be applied to tackle the structural problems of polytypism. The advantages for structure determination are:

— The relevant symmetry relations within a polytypic structure are described by partial coincidence operations (i.e. layer group of the single layer and the symmetry relations between adjacent layers). They allow the interpretation of unusual observations in the diffraction pattern (e.g. symmetry and/or weights of intensities differ for different classes of reflections, systematic absences, which do not refer to any space group aso.), leading in many cases to a first hypothesis of the building principle.

— OD-structures with the same structure principle and differing only in the stacking of the layers belong to a family of OD-structures, the so called OD-family. Thus new relationships between structures are defined, which may be important for structure determination.

— There are special members in any OD-family, OD-structures with maximum degree of order, so called MDO-structures, where the interaction of next but one and more distant neighbours are taken into account. For many polytypic substances periodic polytypes occurring most frequently, turn out to be MDO-Structures.

— The symmetry relations and the stacking of layers in MDO-structures are described by special groupoids of the OD-groupoid family, so called MDO-groupoids, which can be derived, if the OD-groupoid family is known (DORNBERGER-SCHIFF 1982). The knowledge of all MDO groupoids within a family is important for the interpretation of the intensity maxima in the diffraction pattern.

SEDLACEK (1978, 1980) introduced the term of pseudo-lattice structure. This allow a special procedure of structure determination. Pseudo-lattice structures are defined as follows: All structures units of the layers of one sort lie on points of a pseudo-lattice $\boldsymbol{r} = m\hat{\boldsymbol{a}} + n\hat{\boldsymbol{b}} + p\hat{\boldsymbol{c}}$, but only a part of these points isooccupied. The pseudo-lattice is the same for all sorts of layers. A considerable number of polytypes belong to the class of pseudo-lattice structures. In pseudo-lattice structures the $S$ function is periodic. This function is given by

$$S^{(s)} = K_v \sum_{m=0}^{Q_1-1} \sum_{n=0}^{Q_2-1} \sum_{p=0}^{P-1} g^{(s)}_{m,n,p} \varphi^{(s)}_{m,n,p} \tag{IV}$$

with $K_v$ as scale factor

$$\varphi = \exp\left[2\pi i(\boldsymbol{r}_m, \boldsymbol{r}^*)\right] ,$$

$$\boldsymbol{r}^* = h/Q_1\hat{\boldsymbol{a}}^* + k/Q_2\hat{\boldsymbol{b}}^* + \zeta\hat{\boldsymbol{c}}^*$$

and

$$g^{(s)}_{m,n,p} = \begin{cases} 0 \\ 1 \end{cases},$$

where $g^{(s)}_{m,n,p}$ is an occupation factor. The function characterizing the stacking of layers $S^{(s)}$ is periodic as follows from

$$S^{(s)}_{(hkl)} = S^{(s)}_{(h+Q_1\hat{h},\, k+Q_2\hat{k},\, \zeta+\hat{l})} \tag{V}$$

with the relations

$$\hat{\boldsymbol{a}} = \boldsymbol{a}/Q_1 , \qquad \hat{\boldsymbol{b}} = \boldsymbol{b}/Q_2 .$$

The tripel $(\hat{h}, \hat{k}, \hat{l})$ correspond to a superposition structure with the basic vectors $(\hat{\boldsymbol{a}}, \hat{\boldsymbol{b}}, \hat{\boldsymbol{c}})$.

All reflections $h = h_\lambda + Q_1\hat{h}$, $k = k_\lambda + Q_2\hat{k}$, $\zeta = \zeta_\lambda + \hat{l}$, with $(h_\lambda, k_\lambda, \zeta_\lambda) = \text{const.}$ belong to a class of reflections ($\lambda$) with the same $S^{(s)}$ function.

The $K^{(s,t)}$ function is represented according to (IV) as follows

$$K^{(s,t)} = k_1 \sum_{m=0}^{Q_1-1} \sum_{n=0}^{Q_2-1} \sum_{p=-(P-1)}^{(P-1)} N^{(s,t)}_{m,n,p} \varphi_{m,n,p} . \tag{VI}$$

The vectors weights $N^{(s,t)}_{m,n,p}$ are proportional to the number of vectors $(m, n, p)$ between the layers from sort (s) to the sort (t).

The Fourier transformation of $K^{(s,t)}$ is equal to

$$\pi_{m,n,p} = K_2 \sum_{m=0}^{Q_1-1} \sum_{n=0}^{Q_2-1} \sum_{p=-(P-1)}^{(P-1)} N^{(s,t)}_{m,n,p} \delta'(u - m/Q_1, v - n/Q_2, w - p) \tag{VII}$$

with

$$\lim_{p\to\infty} \delta' = \delta ,$$

where $\delta$ is the Dirac function. If the intensities along the diffuse rods are measured at points $\zeta = l/Q_3$ (with $l$ as an integer), the corresponding Fourier transform becomes artificially periodic with $\boldsymbol{c} = Q_3\hat{\boldsymbol{c}}$. Corresponding to this fact, the fictitious $N'^{(s,t)}_{m,n,p}$-values are now superpositions of the real $N^{(s,t)}_{m,n,p}$-values in the fictitious cell $\boldsymbol{a}, \boldsymbol{b}, \boldsymbol{c} = Q_3\hat{\boldsymbol{c}}$. The Fourier transformations of the $K^{(s,t)}$-values of a separate set of reflections ($\lambda$) yield fictitious $N'^{(s,t)}_{m,n,p}$-values, which must be interpreted as additional superpositions, modulated by phase factors (SEDLACEK 1978).

## 2. Indirect methods for structure determination of polytypes

Firstly has been determined the structures of three SiC-polytypes, i.e. the 4H, 6H and the 15R-structure (OTT 1925 a, b, 1926). The result of BAUMHAUER (1912), obtained by morphological measurements has been confirmed. In the following years other polytypic substances have been described, as ZnS, $CdJ_2$, $CdBr_2$, $K_4(FeCN)_6$ and others (BRAGG 1920; BOZORTH 1922; PINSKER 1953); HASER et al. 1969; TRIGUNAYAT, CHADHA 1971). At most the structure determination was based on conventional methods, according to the known methodology.

The structure determination of long-periodic polytypes involves some difficulties. The number of possible different structures increases rapidly with the number of layers per repeat unit. Mathematical methods for calculation of the number of different polytypes, as given by MOGAMI et al. (1978), IGLESIAS (1981) and MCLARNAN (1981) are to complicated and not useful for the description of the real relations. In reality there is only a minor number of polytypes realized. For the practical structure determination this means, that it is necessary to develope methods for an extreme decrease of possible structures. One of these is the MDO-philosophy given by DORNBERGER-SCHIFF.

Further difficulties result from the so called "syntactic coalescence" of polytypes. There are some regions in one and the same crystal with different kinds of stacking. Therefore the different sets of intensities belonging to the different kinds of stacking must be separated. This is possible by experimental methods or in a later phase of structure analysis.

In the first stage of structure determination of polytypes "trial and error"-methods have been used. These methods are well suited for investigation of short-periodic polytypes. The difficulties mentioned above gave rise for the development of methods for the investigation of long-periodic polytypes. With the progress of computer technique it is possible to simulate the intensity distribution of such a structure. The real structure results from a direct or computational comparison of the intensity distribution. But if there is a high number of possible structures, this procedure is very computer-time consuming.

For high-order polytypes the intensity distribution function of all possible structures have been published by MARDIX et al. (1970). In order to reduce the computer time, preliminary informations has been used, which simultaneously reduce the number of possible structures. These informations are:

— the number of layers per repeat unit (from oscillation photographs),
— the polytype being rhombohedral or not (from extinctions),
— the cyclycity of the polytype (by optical measurements) (BRAFMAN, STEINBERGER 1966).

These informations are used to form all possible Zhdanov symbols. A computer program for calculating the intensities according to the possible Zhdanov symbols eleminates all sets of intensities, which are not compatible with the observed intensities. It is a high-speed method for the determination of high-order polytypes and has been applied to more than fifty ZnS-polytypes (MARDIX et al. 1970).

A further concept to facilitate the structure determination is that of the so called "polytype family". A polytype family is defined as consisting of all structures with the same periodicity (m), as well as of the rhombohedral structure with 3m layer periodicity. This concept has been published by MARDIX et al. (1967).

KRISHNA and VERMA (FARKAS-JAHNKE) have used systematical non-space group extinctions for the analysis of $36H_b$-polytype of SiC. These extinctions caused by the stacking function $S$ are observed for intensities $(10l)$ with $l = q$, $q = (16n, 36n + 6, 6n + 2)$. Therefore $S(10l)$ is equal to

$$S(10l) = \sum_{m=0}^{2} \sum_{p=0}^{35} g_{m,p} \exp\left[2\pi i \left(\frac{m}{3} + q\frac{p}{36}\right)\right] = 0 . \qquad \text{(VIII)}$$

With these relations the structure was determined and systematic extinctions were interpreted as caused by partially periodic regions of the crystal.

#### References


ALLEGRA, G.: Acta Cryst. **14**, 535 (1961)
ALLEGRA, G.: Acta Cryst. **17**, 579 (1964)
BAILEY, S. W. et al. a): Acta Cryst. **A33**, 681 (1977)
BAILEY, S. W. et al. b): Amer. Miner. **62**, 411 (1977)
BAUMHAUER, H.: Z. Krist. **50**, 33 (1912)
BOZORTH, R. M.: J. Amer. Chem. Soc. **44**, 2232 (1922)
BRAFMAN, O., STEINBERGER, I. T.: Phys. Rev. **143**, 501 (1966)
BRAGG, W. L.: Phil. Mag. **39**, 647 (1920)
DORNBERGER-SCHIFF, K., GRELL-NIEMANN, H.: Acta Cryst. **14**, 167 (1961)
DORNBERGER-SCHIFF, K.: Abh. Dtsch. Akad. Wiss., Kl. f. Chemie **3** (1964)
DORNBERGER-SCHIFF, K.: Lehrgang über OD-Strukturen, Akademie-Verlag, Berlin (1966)
DORNBERGER-SCHIFF, K.: Kristall und Technik **14**, 1027 (1979)
DORNBERGER-SCHIFF, K.: Polytypes and MDO-Structures, Collected Abstracts "symposium on special topics of X-ray crystal structure analysis", Hohengrün 1980
DORNBERGER-SCHIFF, K.: Acta Cryst. **A38**, 483 (1982)
FARKAS-JAHNKE, M.: Acta Cryst. **18**, 571 (1965)
GEVERS, R.: Acta Cryst. **5**, 518 (1952)
GEVERS, R. a): Acta Cryst. **7**, 337 (1954)
GEVERS, R. b): Acta Cryst. **7**, 492 (1954)
HASER, R. et al.: C.R. Acad. Sci. (France) **268**, 51 (1969)
HENDRICKS, S. B., TELLER, E.: J. Chem. Phys. **10**, 147 (1942)
IGLESIAS, J. E.: Z. Krist. **155**, 121 (1981)
JAGODZINSKI, H. a): Acta Cryst. **2**, 201 (1949)
JAGODZINSKI, H. b): Acta Cryst. **2**, 208 (1949)
JAGODZINSKI, H. c): Acta Cryst. **2**, 298 (1949)
JAGODZINSKI, H.: Acta Cryst. **7**, 17 (1954)
JOHNSON, C. A.: Acta Cryst. **16**, 490 (1963)
KAKINOKI, J., KOMURA, Y.: J. Phys. Soc. Jap. **7**, 30 (1952)
KAKINOKI, J., KOMURA, Y. a): J. Phys. Soc. Jap. **9**, 169 (1954)
KAKINOKI, J., KOMURA, 1. b): J. Phys. Soc. Jap. **9**, 177 (1954)
KAKINOKI, J., KOMURA, Y.: Acta Cryst. **15**, 292 (1962)
KAKINOKI, J., KOMURA, Y.: Acta Cryst. **19**, 137 (1965)
KAKINOKI, J.: Acta Cryst. **23**, 875 (1967)
LELE, S. et al.: Acta Cryst. **A25**, 471 (1969)
LELE, S.: Acta Cryst. **A25**, 551 (1969)
MARDIX, S. et al. a): Acta Cryst. **A26**, 599 (1970)
MARDIX, S. et al. b): Acta Cryst. **B26**, 24 (1970)
MARDIX, S. et al.: Acta Cryst. **22**, 805 (1967)
MCLARNAN, T. J.: Z. Krist. **155**, 269 (1981)
MEGAW, M. D.: Proc. Roy. Soc. **A259**, 59 (1960)
MERING, J.: Acta Cryst. **2**, 371 (1949)
MOGAMI, K. et al.: Can. Miner. **16**, 427 (1978)
ONODA, M., KAWADA, I.: Acta Cryst. **A36**, 134 (1980)
ONODA, M. et al.: Acta Cryst. **A36**, 952 (1980)
OTT, H. a): Z. Krist. **61**, 515 (1925)
OTT, H. b): Z. Krist. **62**, 201 (1925)
OTT, H.: Z. Krist. **63**, 1 (1926)
PATTERSON, M. S.: J. Appl. Physics **23**, 805 (1952)
PINSKER, Z. G.: London 1953, p. 271
PRASAD, B., LELE, S.: Acta Cryst. **A27**, 54 (1971)
SATO, R.: Acta Cryst. **20**, 150 (1966)
SEDLACEK, P.: Dr. sc. Thesis, Akademie der Wissenschaften der DDR, Berlin 1978
SEDLACEK, P.: Collected Abstracts "Symposium on special topics of X-ray structure analysis", Hohengrün 1980
TAKAGI, Y. et al.: Acta Cryst. **A32**, 657 (1976)

TAKAKI, Y.: Acta Cryst. **A33**, 690 (1977)
TRIGUNAYAT, G. C., CHADHA, G. K.: phys. stat. sol. (a) **4**, 9 (1971)
VAN DER PERRE, G., DERUYTTERE, A.: J. Appl. Cryst. **13**, 65 (1980)
WARREN, B. E. et al.: Acta Metallurg. **3**, 473 (1955)
WARREN, B. E.: Prog. Met. Phys. 8, 147 (1959)
WILSON, A. J. C.: Proc. Roy. Soc. A, **180**, 277 (1942)
WILSON, A. J. C.: Proc. Roy. Soc. A, **181**, 360 (1943)
ZACHARIASEN, W. H.: Phys. Rev. **71**, 715 (1947)

(Received 1) October 17, 1986;
2) December 3, 1986)

*Authors' address*:

c/o Dr. R.-J. KUBAN
Zentralinstitut für physikalische Chemie
der Akademie der Wissenschaften der DDR
Rudower Chaussee 5
Berlin
DDR-1199

Cryst. Res. Technol. | 22 | 1987 | 6 | 799–802

R.-J. KUBAN, B. SCHULZ

Zentralinstitut für physikalische Chemie,
Akademie der Wissenschaften der DDR, Berlin-Adlershof

# Crystal Structure Determination of 2-Acetylamino-4-benzoyl-5,5-dimethyl-4,5-dihydro-1,3,4-thiadiazolin, $C_{13}H_{15}N_3O_2S$

Thiosemicarbazone is receiving increasing interest because of the biological activity of some of its derivatives. Mr = 277.35, space group $P2_12_12_1$, $Z = 4$, $a = 9.290(3)$, $b = 11.512(2)$, $c = 13.286(2)$ Å, $V = 1421(1)$ Å$^3$, $D_m = 1.29(1)$ Mg m$^{-3}$, $D_x = 1.296$ Mg m$^{-3}$, $\lambda(\text{MoK}_\alpha) = 0.71062$ Å, $\mu = 0.230$ mm$^{-1}$, $F(000) = 584$, $T = 296$ K, final $R = 0.028$ for 1255 reflections. All hydrogen atoms were located. The structure determed by X-ray analysis confirms the chemical results. The structure analysis of the title compound is described and its molecular and crystal structure discussed.

Das Interesse am Thiosemicarbazon und dessen Derivate ist infolge der biologischen Aktivität dieser Verbindungen gestiegen. Mr = 277.35, Raumgruppe $P2_12_12_1$, $Z = 4$, $a = 9.290(3)$, $b = 11.512(2)$, $c = 13.286(2)$ Å, $V = 1421(1)$ Å$^3$, $D_m = 1.29(1)$ Mg m$^{-3}$, $D_x = 1.296$ Mg m$^{-3}$, $\lambda(\text{MoK}_\alpha) = 0.71062$ Å, $\mu = 0.230$ mm$^{-1}$, $F(000) = 584$, $T = 296$ K, $R = 0.028$ für 1255 Reflexe. Alle Wasserstoffatome wurden lokalisiert. Die Ergebnisse der Röntgenstrukturanalyse stimmen mit den chemischen Ergebnissen überein. Die Strukturanalyse der oben genannten Verbindung wird beschrieben und die Molekül- und Kristallstruktur werden diskutiert.

## Introduction

Thiosemicarbazone and thiosemicarbazone derivatives have been studied because of their pharmacological activity against tuberculosis (DOMAGK et al.). They have also been found to be active against certain kinds of tumors (PETERING et al.), smallpox and vaccinia (BAUER, SADLER), and have been suggested as possible fungicides (BENNS

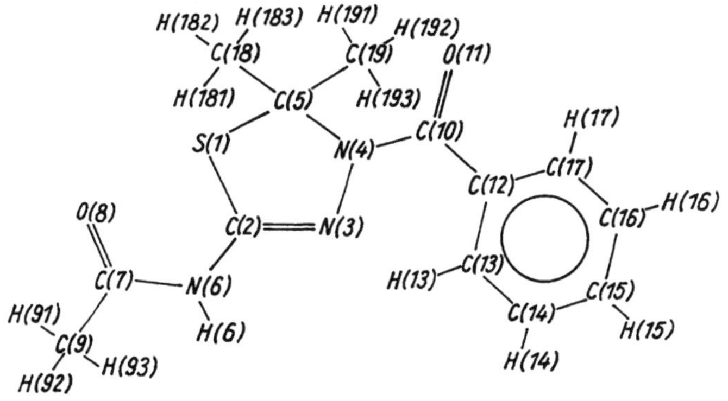


Fig. 1. Structural formula of ABDT with numbering

et al.). In order to gain a better knowledge of this heterocycle we have carried out X-ray investigations of crystal structures containing the 2-acetylamino-4-benzoyl-5,5-dimethyl-4,5-H-1,3,4-thiadiazolin (for short ABDT). The shape of the molecule is indicated in Fig. 1.

## Experimental

ABDT was prepared from the thiosemicarbazides by acylation, following the procedure of ANDREAE et al. $D_m$ from flotation method (benzene-carbon tetrachloride). Prismatic crystals $0.12 \times 0.15 \times 0.25$ mm³; preliminary determination of lattice parameters and crystal symmetry (space group $P2_12_12_1$ from systematic absences: $h00$, $h = 2n$; $0k0$, $k = 2n$ and $00l$, $l = 2n$) made with a Weissenberg goniometer. An Enraf-Nonius CAD 4 diffractometer has been used. Cell constants were determined by least-squares using 25 reflections with $3 < \theta < 15°$, graphite monochromator, hemisphere up to $2\theta_{max} = 55°$, $(\sin \theta/\lambda)_{max} = 0.65$ Å$^{-1}$; range of $hkl$: $h0 \rightarrow 12$, $k0 \rightarrow 14$, $l0 \rightarrow 17$, total number of reflections 1378, $\omega - 2\theta$ scan, three standard reflections every 3000 seconds remained stable; Lorentz-polarization correction, no absorption correction, 1255 reflections with $I > 1\sigma(I)$ were used, weighting scheme $w(I) = 1/\sigma^2$; structure determination by MULTAN 11/82 (MAIN et al.) using $E > 1.39$, H-atom positions determined from electron density difference map, full-matrix least-squares procedure minimized $\Sigma w(\Delta F)^2$, 233 parameters were refined, extinction parameters included, unweighted $R$ factor $= 0.0292$, unweighted $R$ factor including unobserved reflections 0.033, final difference synthesis peaks $< 0.2e$ Å$^{-3}$, largest parameter shift $= 0.003$, scattering factors from International Tables for X-ray Crystallography (1974), program system applied: Enraf-Nonius SDP-PLUS version 1.0

## Discussion

The final fractional coordinates and the corresponding standard deviations of the atoms are listed in Table 1. A drawing of the ABDT molecule, including bond distances

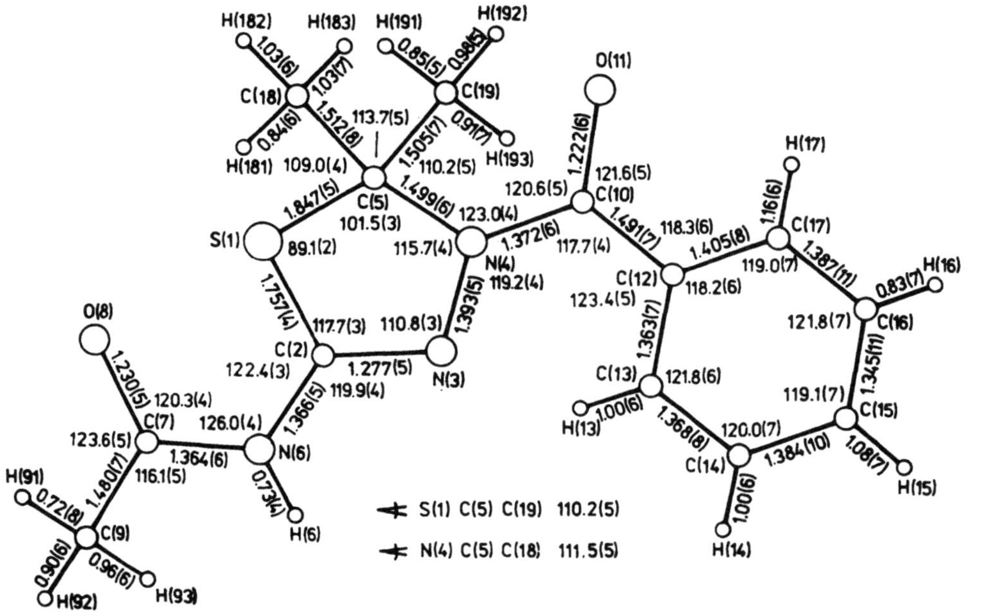

Fig. 2. The intramolecular bond lengths with estimated standard deviations in parentheses

Table 1

Fractional coordinates ($\times 10^4$) and mean temperature factors for non-hydrogen atoms. E.s.d.'s are given in parentheses. $U_{eq} = \frac{1}{3} \sum_i \sum_j U_{ij} a_i a_j \underline{a}_i \underline{a}_j$

| | $x$ | $y$ | $z$ | $U_{eq}$/Å$^2$ |
|---|---|---|---|---|
| S(1) | 4301.9(8) | 537.8(7) | 6600.0(7) | 0.0504 |
| O(8) | 3696(2) | 1815(2) | 4913(2) | 0.0594 |
| O(11) | 7332(3) | −139(2) | 9223(2) | 0.0821 |
| N(3) | 6857(3) | 1371(2) | 6972(2) | 0.0414 |
| N(4) | 6491(3) | 706(2) | 7810(2) | 0.0491 |
| N(6) | 5987(2) | 1974(2) | 5448(2) | 0.0386 |
| C(2) | 5849(3) | 1352(2) | 6319(2) | 0.0356 |
| C(5) | 5276(4) | −134(3) | 7662(3) | 0.0514 |
| C(7) | 4923(3) | 2174(3) | 4761(2) | 0.0447 |
| C(9) | 5349(4) | 2831(4) | 3851(3) | 0.0681 |
| C(10) | 7436(4) | 638(3) | 8601(2) | 0.0570 |
| C(12) | 8533(3) | 1577(3) | 8702(2) | 0.0526 |
| C(13) | 8230(4) | 2724(3) | 8566(3) | 0.0650 |
| C(14) | 9220(5) | 3572(4) | 8769(3) | 0.0823 |
| C(15) | 10592(4) | 3273(4) | 9082(3) | 0.0937 |
| C(16) | 10921(4) | 2144(5) | 9202(3) | 0.0988 |
| C(17) | 9918(4) | 1272(4) | 9035(3) | 0.0810 |
| C(18) | 4296(4) | −169(3) | 8570(3) | 0.0771 |
| C(19) | 5852(4) | −1305(3) | 7361(3) | 0.0751 |

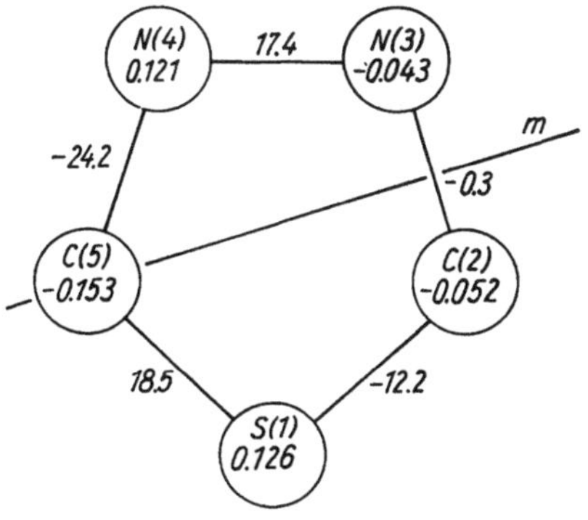


**Fig. 3. The puckering of the five-membered ring in ABDT. In the circles the numerical values are the perpendicular displacements (Å), of the corresponding atoms of the mean plane through the ring. (Torsion angles in degrees.)**

and angles, is shown in Figure 2. The bond distances and angles in the ABHT molecule are comparable to those observed in several thiophene derivatives.

There is a 1,3,4-thiadiazoline ring in the molecule: S(1)C(2)N(3)N(4)C(5). It is evident that it cannot be planar. In particular, the puckering amplitude $Q$, which defines quantitatively the degree of puckering (Cremer, Pople) is 0.24(1) Å. Within ring S(1)C(2)N(3)N(4)C(5) there is a pseudo mirror plane through C(5) and the mid-point of the C(2)—N(3). Figure 3 and the asymmetry parameter of Duax et al. (1976), $\Delta C_s$ C(5) = 5.5°, indicate that the ring assumes an envelope conformation. The phenyl ring C(12)—C(17) is plane within the limits of experimental error.

The individual five- and six-membered rings are not coplanar, having a dihedral angle of 48.1° between the thiodiazoline and benzene rings.

The packing of the moleculs in the unit cell is shown in Figure 4. There is a short-N–H ... O contact involving N(6), which is 2.917(5) Å from O(8) of an adjacent molecule ($\frac{1}{2} + x$; $\frac{1}{2} - y$; $1 - z$). The H(6) ... O(8) distance of 2.19(4) Å together with the

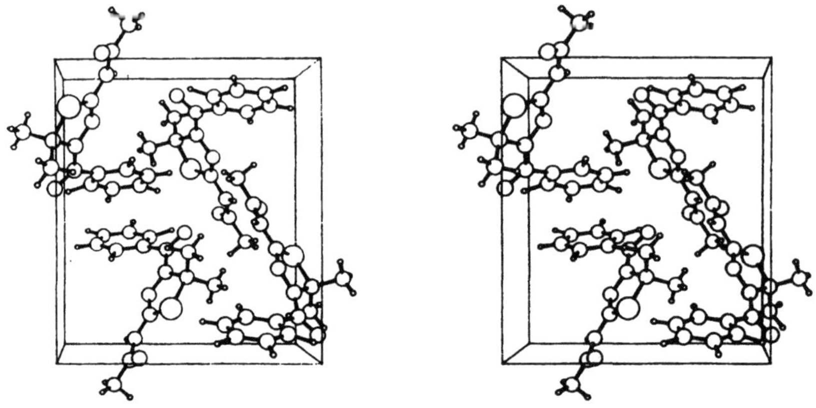

Fig. 4. Packing of molecules in the crystal. This is a view down the $a$ axis of the unit cell

N(6)–H(6) ... O(8) angle of 178.25° suggest that a weak N–H ... O bond may be present. Packing is due mainly to van der Waals interactions and all other intermolecular contacts agree with those predicted from radii-sum rules.

The authors thank Dr. S. Andreae for crystals.

**References**

ANDREAE, S., SCHMITZ, E., SEEBOTH, H.: J. prakt. Chem., submitted for publication

BAUER, D. J., SALDLER, P. W.: Chem. Astr. **65**, 8683b (1966)

BENNS, B. G., GINGRAS, B. A., BAYLEY, C. H.: Appl. Microbiol. 8, 353 (1960)

CREMER, D., POPLE, J. A.: J. Amer. Chem. Soc. **97**, 1354 (1975)

DOMAGK, G., BEHNISCH, R., MIETZSCH, F., SCHMIDT, H.: Naturwissenschaften **33**, 315 (1946)

DUAX, W. L., WEEKS, C. M., ROHRER, D. C.: Stereochemistry, Vol. 9, New York 1976

International Tables for X-ray Crystallography. Vol. IV, Birmingham 1974

MAIN, P., FRISKE, S. J., HULL, S. E., LESSINGER, L., GERMAIN, G., DEQLERCG, J.-P., WOLFSON, M. M.: MULTAN 11/82. A System of Computer Programs for the Automatic Solution of Crystal Structures from X-ray Diffraction Data. Univ. of York, England, and Louvain, Belgium 1982

PETERING, H. G., BUSKIRK, H. H., UNDERWOOD, G. E.: Cancer Res. **24**, 367 (1964)

(Received November 1, 1986)

*Authors' adress:*

c/o Dr. Ralf-Jürgen KUBAN
Zentralinstitut für physikalische Chemie
der Akademie der Wissenschaften der DDR
Rudower Chaussee
Berlin-Adlershof
DDR-1199

| Cryst. Res. Technol. | 22 | 1987 | 6 | 803–807 |
| --- | --- | --- | --- | --- |

C. C. Desai, A. N. Hanchinal

Department of Physics, Sardar Patel University, Vallabh Vidyanagar, Gujarat, India

# Microstructures of Gel-grown Ammonium Hydrogen Tartrate Single Crystals

Single crystals of ammonium hydrogen tartrate (AHT) have been grown in silica gels by employing the controlled reaction between ammonium chloride and tartaric acid. Transparent AHT crystals upto $24 \times 4 \times 3$ mm$^3$ in size have been grown at room temperature. Optical and electron-optical studies have been made on the various surface structures of $\{010\}$ faces of the grown crystals. A variety of growth striations and growth hillocks have been observed. Growth layers modified by the presence of misaligned microcrystals have been illustrated. It has been suggested that two-dimensional nucleation, spreading and pilling up of growth layers is mainly responsible for the growth of these crystals and the implications are discussed.

## 1. Introduction

Microtopographical studies of crystals contribute a great deal to the understanding of the mechanism of the growth of the crystals. Although the surface structures of different faces of crystal represent the conditions at the last stage of growth, correlation of such features of different faces lead, to a great deal to the history of growth of the crystals. Since imperfections affect the growth and dissolution pattern, one can also obtain information regarding these. Various investigators like Seager; Verma; Tolansky; Augustine, Hale; Sunagawa; Desai; Desai, Rai, etc. have made attempt to unfold the growth mechanism of various crystals by studying their growth patterns.

Ammonium Hydrogen Tartrate (AHT) crystals are of considerable interest, particular for their basic studies of their ferroelectric, piezoelectric, optical and other pertient characteristics. They are extensively used for controlling laser emission. The efficiency of AHT crystals in different modulation and defocusing systems are characterized by their electro-optical coefficients (Pipree, Koblova). In the present investigation, single crystals of AHT have been grown by gel method (Desai, Hanchinal) and a detailed investigation on microstructures on the habit faces of these crystals have been made with specific aim of understanding the mechanism of its growth.

## 2. Experimental procedure

AHT crystals were grown at constant temperature of 30 °C at $p$H between 4 to 5 using 1.04 gm cm$^{-3}$ specific gravity of sodium metasilicate. They were grown using various concentrations of feed solutions varying from 1.0 M to 5.0 M. The crystals were thoroughly cleaned with ether and were coated with silver film in a vacuum coating unit to enhance the contrast. These faces were then examined under a metallurgical and electron microscopes.

## 3. Observations and discussion

The grown crystals were examined by scanning electron microprobe analysis, X-ray diffraction, density measurement and thermogravimetric analysis, these techniques confirmed that they were AHT crystals. Typical crystals of AHT grown in the present work are usually transparent and in the form of long needles. Crystals in the habit of rod-like, platelet and orthorhombic disphenoidal were also obtained.

### 3.1. Growth layers from the edges or high steps on the crystals

The edges of the crystal or high steps on the crystal surface sometimes work as elegant centers for initiation of growth layers. Several examples illustrating this fact were observed on the (010) faces of the grown crystals. A typical example is shown in Figure 1, wherein the starting of growth layers from several points on the dark line

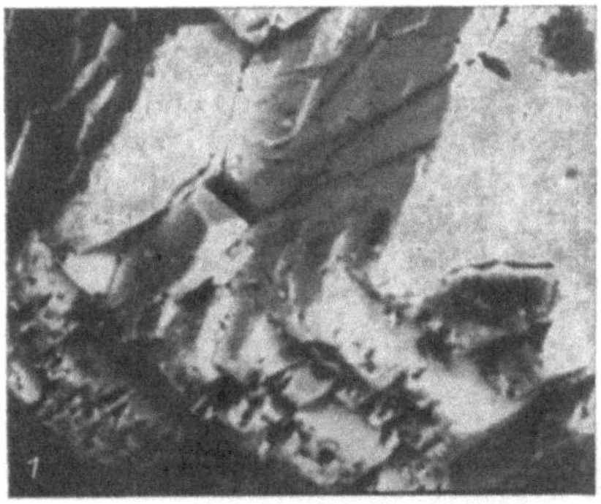

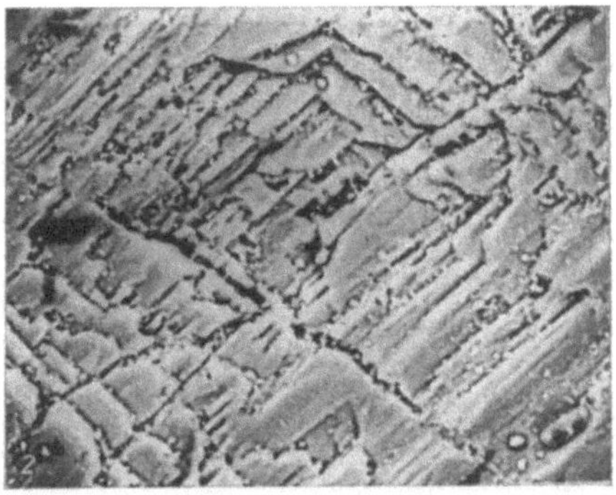

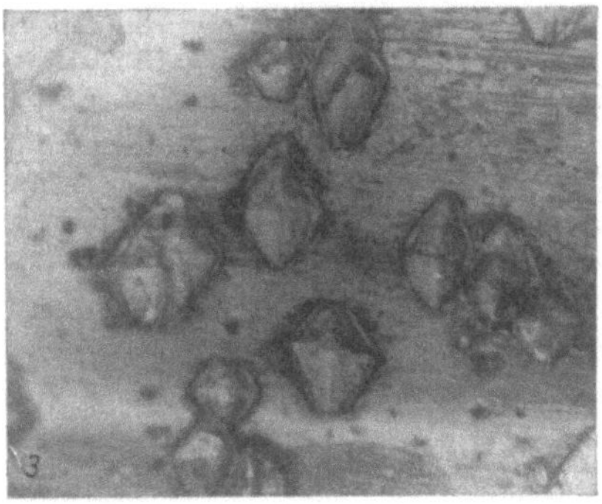

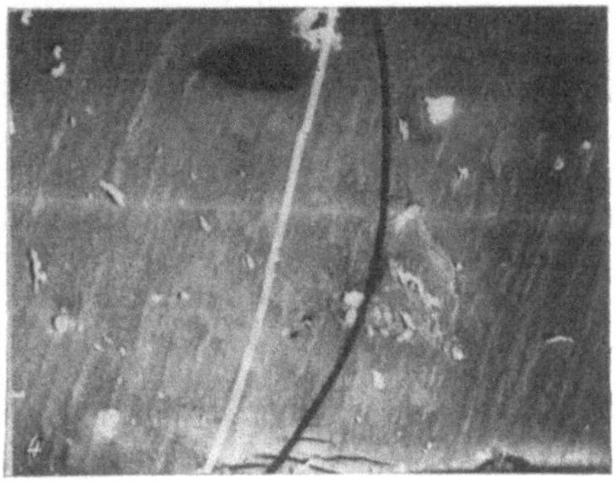

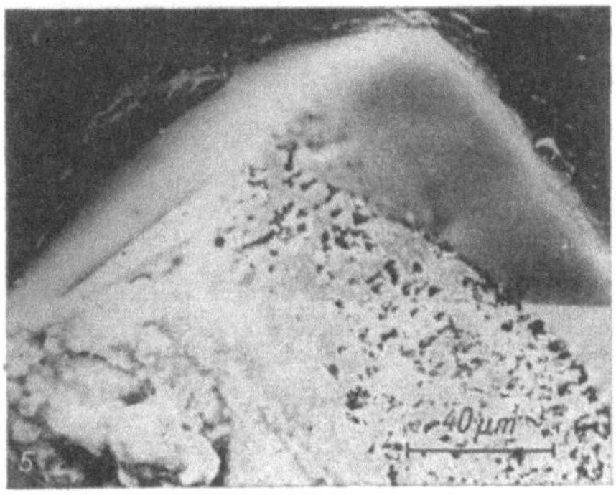


Fig. 1. Growth layers initiating from the edge of a crystal
Fig. 2. Initiation of growth layers from step imperfections
Fig. 3. Microcrystals are attached to (100) host face
Fig. 4. Microcrystals formed on (010) surface
Fig. 5. Liquid inclusions with no crystallographic orientation

marked AB can be very clearly seen. This line marked AB is the edge of thick crystal on which the growth layers are seen. This figure shows a build up of material on the crystal edges with a fillin towards the centre indicating that edge nucleation was the predominant growth mechanism at higher concentrated feed solution.

The occurrence of step imperfections is a common feature observed on crystals grown with 2.0 M feed solutions and often layers do originate from the imperfections on the steps. A beautiful example is provided in Figure 2 where step imperfections initiate growth layers.

### 3.2. Growth of microcrystallites or attaching of secondary crystals

Sometimes guest crystals of quite a large size (from a few angstroms to a few microns) were observed to the (010) faces of the host (primary) crystals. The attached faces of the secondary crystals may be {100}, {110} or {130}. Figure 3 shows a case in which {100} faces of microcrystals are attached to the host face. The guest crystals are strictly crystallographically oriented. The secondary crystals might have formed while the crystals were being removed from the gel. Since the feed solutions contained both types of ions ($NH_3^+$ and $Cl^-$), as soon as the gel was removed from the crystallization apparatus, these two types of ions might have come together to form secondary crystallites. According to BUCKLEY overgrowth occurs owing to interruptions in the continuity of deposition due to changes in growth conditions, such as temperature changes or the temperary exhaustionof a solution after an active burst of crystallization. Such overgrowth is shown in Figure 4.

### 3.3. Liquid inclusions

Inclusions are produced when an area of a crystal face ceases to grow whilst the neighbouring areas continue to grow normally, then after a certain time, the available liquid is enclosed the reason of the cessation of growth may be adsorption of some impurity, or some minor fluctuations in temperature during growth. The probable liquids trapped into the crystals are, in the authors' opinion, HCl resulting as the by-product or reaction waste product during the crystallization of AHT. Liquid inclusion with no crystallographic orientation have been observed as shown in Figure 5.

### 3.4. Growth hillocks

Prism faces (111) and ($\bar{1}\bar{1}\bar{1}$) in disphenoidal tetrahedral crystals exhibited on them densely populated oriented growth hillocks. Figure 6 illustrates a region of a face with such hillocks. The growth layers composing the hillocks are distinctly seen. The photograph also shows interaction of growth fronts of layers extending from neighbouring growth hillocks. Their longer dimensions are parallel to the $c$-axis of the crystal.

### 3.5. Dissolution lines an growth striations

Interestingly dissolution lines are observed on the (010) surfaces of the crystals grown at high temperatures between 35° to 45 °C. This may be because of the increase in the solubility of the crystal with temperature, such dissolution lines are illustrated in Figure 7. Sometimes grown striations are also observed as shown in Figure 8. These are believed to be due to self exciting variations in growth rate, that arose from the temperature oscillations.

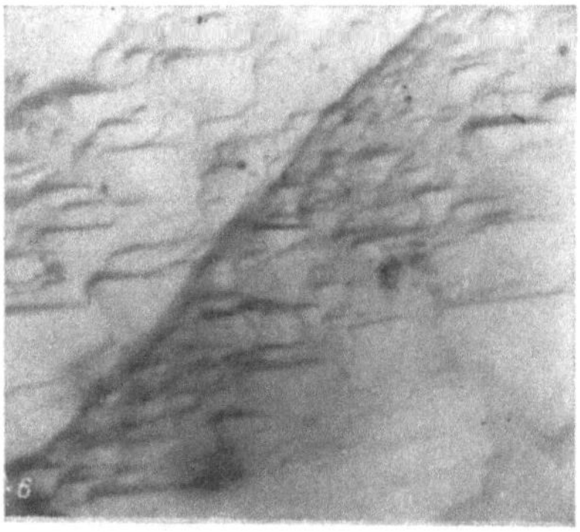

Fig. 6. Triangular shaped growth hillocks on (010) surfaces
Fig. 7. Crystallographic oriented dissolution lines on (010) surface
Fig. 8. Growth striations on the as-grown (010) surface

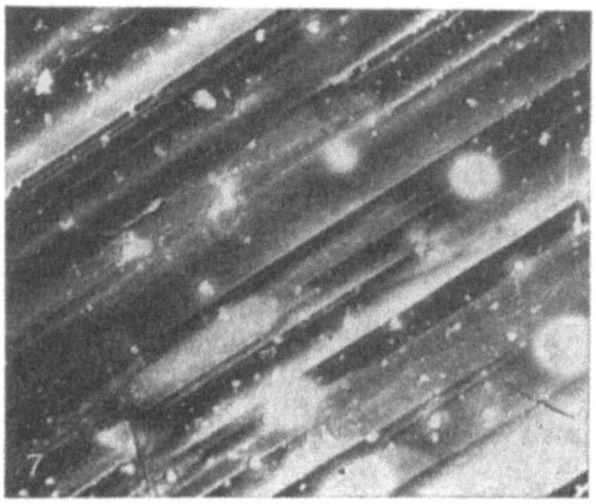

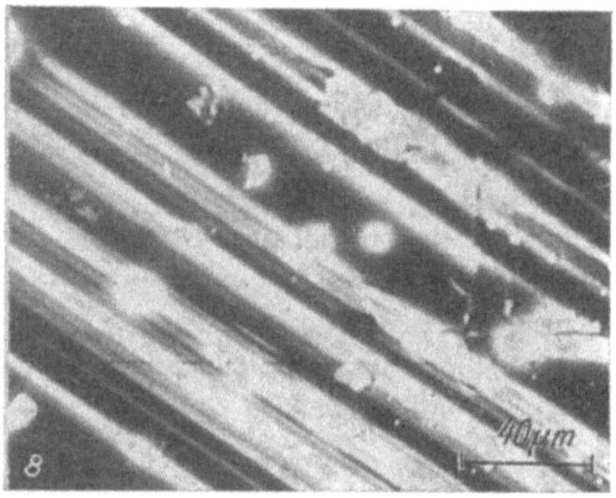


## 4. Conclusions

1. At all concentrations of feed solutions layer growth mechanism is predominant. It has been observed that the growth layers initiate from the edges, steps and inclusions on the crystal surfaces.
2. Microcrystallites with crystallographic orientation have been observed on the host crystal faces.
3. It has been observed that the liquid entrapment is the main source of crystal defects in gel grown AHT crystals.
4. Strictly crystallographically oriented hillocks on (010) surfaces are developed by a layer mechanism.
5. Advancing growth layers might produce guest crystals.
6. Dendrite patterns spread on the faces are due to solute solidifcation at the end stage of growth.

One of us (CCD) thank University Grants Commission (UGC), New Delhi for the award of the project entitled "Studies of growth and characterization and properties of alkali halides and tatrate single crystals".

## References

Augustin, M., Hale, D. R.: J. Phys. Chem. Solids **13**, 344 (1960)
Buckley, H. E.: Crystal Growth, New York (1951)
Desai, C. C.: Kristall und Technik **14**, 289 (1979)
Desai, C. C., Hanchinal, A. N.: Cryst. Res. Technol. **16**, 1231 (1981)

DESAI, C. C., RAI, J. L.: Cryst. Res. Technol. **16**, 1231 (1981)
PIPREE, L. V., KOBLOVA, M. M.: Radio Eng. Electronic Physics (USA) **12**, 503 (1967)
SEAGER, A. F.: Miner. Mag. **30**, 1 (1953)
SUNAGAWA, I.: Bull. Geol. Sun. Japan **2**, 129 (1951)
TOLANSKY, S.: Proc. Roy. Soc. **A 270**, 443 (1962)
VERMA, A. R.: Phil. Mag. **42**, 1005 (1951)

(Received November 11, 1986)

*Authors Addresses*:

Professor C. C. DESAI
Department of Physics, Sardar Patel University
Vallabh Vidyanagar 388120
Gujarat, India

Dr. A. N. HANCHINAL
Indian Institute of Geomagnetism
Colaba
Bombay 400005
India

**Book Review**

J. S. Blakemore

## Solid State Physics

Cambridge University Press 1985, 2. Auflage, Preis: £ 40.00, ISBN 0-521-30932-8.

Das Buch erschien 1969, die zweite Auflage 1974. Die vorliegende Ausgabe ist die verbesserte zweite Auflage. Der Inhalt des Buches entspricht dem einer einsemestrigen Vorlesung über Festkörperphysik mit der Betonung der Elektronenstruktur des Festkörpers. Der Autor setzt Kenntnisse der Quantentheorie und Elektrodynamik voraus. In dem Buch werden die theoretischen Konzepte erläutert, die zur Beschreibung der Schwingungsspektren, der Elektronen-Energiebänder sowie der elektrischen, dielektrischen und magnetischen Eigenschaften von Festkörpern verwendet werden. Die Eigenschaften periodischer Strukturen stehen dabei im Vordergrund. Das Buch ist jedoch keinesfalls rein theoretisch orientiert. Bei vielen Problemen wird der theoretischen Ableitung die Veranschaulichung vorgezogen, und immer werden Meßergebnisse aus der Originalliteratur zur Untermauerung der Aussagen herangezogen. Auch wird versucht, eine Brücke zu schlagen, wenn sich die Darstellungen in der experimentell bzw. theoretisch orientierten Literatur unterscheiden. So werden Beugungserscheinungen sowohl mit Wellenlängen und Millerschen Indizes wie auch mit Wellenvektoren beschrieben.

Hervorzuheben sind die zusammengefaßte Diskussion der Beugung von Röntgenstrahlen, Elektronen und Neutronen, die detaillierte Diskussion der elektrischen Eigenschaften von Halbleitern und die bis zum Erfassen der Hyperfeinstruktur gehende Darstellung der Elektronenspinresonanz. Es wird ein sehr guter Überblick über die elektrischen Eigenschaften der Metalle gegeben. Jedoch wird es für nicht vorteilhaft angesehen, daß die Elektronentheorie der Metalle zunächst unter Verwendung der Boltzmannstatistik diskutiert wird (3.2).

Es bleibt nicht aus, daß bei einer qualitativen Diskussion ungeeignete Veranschaulichungen auftreten bzw. mit Begriffen operiert wird, die nicht erklärt worden sind. So wird z. B. der Eindruck erweckt, als seien Streuprozesse, bei denen sich der Streuvektor um einen Vektor des reziproken Gitters unterscheidet, äquivalent (2—19, 2—79). In (2—2) wird die Gruppengeschwindigkeit nicht von der Phasengeschwindigkeit unterschieden. Die Betonung der k-Abhängigkeit der Zustandsdichte im k-Raum (S. 100) führt zu falschen Vorstellungen. Wesentlich ist die konstante Zustandsdichte im k-Raum entsprechend der Quantelung des Phasenraumes. Die Begründung, warum das Bild fast freier Elektronen erfolgreich verwendet werden kann (S. 156, 230) ist unvollständig bzw. falsch. Bei der Diskussion der Quantelung im Magnetfeld sollte man sich mit einer rein qualitativen Diskussion begnügen oder die Schrödingergleichung mit Vektorpotential ausführlich erklären (3—171). Schließlich verwundert es, daß der Stern-Gerlach-Versuch auf 1922 datiert wird (S. 445) und Goudsmid und Uhlenbeck zur Erklärung der dort gefundenen Aufspaltung 1925 die Idee des Elektronenspins entwickelt haben sollen.

Diese kritischen Bemerkungen ändern nichts daran, daß das Buch zum Studium der Elektronenstruktur von Festkörpern sehr empfohlen werden kann. Die zahlreichen Übungsaufgaben am Ende jedes Kapitels geben Gelegenheit, den Stoff zu festigen und vertiefende Studien zu betreiben.

P. Rennert

| Cryst. Res. Technol. | 22 | 1987 | 6 | 809–815 |
|---|---|---|---|---|

C. C. DESAI, M. S. V. RAMANA

Department of Physics, Sardar Patel University
Vallabh Vidyanagar, Gujarat, India

# Electrical Conduction in Ferroelectric Lead Hydrogen Phosphate Single Crystals

Single crystals of LHP (Lead Hydrogen Phosphate) have been grown using the controlled reaction between lead nitrate and orthophosphoric acid solution by diffusion process in silica gel medium. Transparent crystals upto $6 \times 4 \times 3$ mm$^3$ in size have been grown at room temperature. The d.c. electrical conductivity of both single crystals and pelletized samples of LHP have been studied in the temperature range from 313 to 968 K. It has been observed that: (i) the d.c. conductivity of the crystalline sample is greater than that of thepellet sample, (ii) three distinct electrical conductivity processes have been observed and are interpreted as extrinsic, intrinsic, phase changes, and (iii) the activation energy for the conduction in the crystalline sample is greater than that of the pellet sample. The magnetic susceptibility measurements reveal that the paramagnetic characteristic of these materials and the implications are discussed.

## 1. Introduction

The electrical conductivity is one of the experimental tools to probe into the structural defects in the crystalline materials. Eventhough the available literature supplies information about the electrical conductivity of alkali halides (ESHEBY et al.; DREYFUS, NOWICK; WARRIS; LAL, DAR; DESAI, RAI) and rare earth tungstate such $EuWO_4$ (RIGDON, GRACE). The electrical conductivity being explained in terms of electrons (or holes), polarons and impurities. A meagre literature is recorded about schultenite crystals such as lead hydrogen phosphate (LHP), $PbHPO_4$. This crystal represents a noval family of ferroelectrics with hydrogen (deuterium bonded structures and which are characterized by one dimension chain of $(PO_4)^{3-}$ tetrahedra strong bonded with protons. SCHULTEN as well BREZINA and HAVRANKOVA have grown platelet crystals of LHP, $PbHAsO_4$ and $PbDAsO_4$ by using water solution method.

Schultenite crystals are of considerable interest, particular for their basic studies of their ferroelectric, piezoelectric, optical and other pertinent characteristics. They are used in transducers and in many linear and non-linear mechanical devices. They are also apparently suitable for testing microscopic theory of ferroelectricity. As far as the authors are aware electrical conductivity measurements of LHP crystals have not been reported in the literature. In this paper, we report on the growth of single crystals of LHP by using gel technique and measurements of their electrical conductivity at different temperatures and its magnetic susceptibility. An attempt has been made to understand the mechanism of charge transport.

## 2. Experimental

### 2.1. Crystal growth technique

Single crystals of LHP used in the present investigations were grown by employing the gel method (DESAI, RAMANA). The growth process involves the diffusion of lead nitrate

Fig. 1. Typical crystals of LHP grown in the present work

into a gel in which orthophosphoric acid is dissolved before hand. The general chemical reaction taking place to give the required LHP crystals are as follows:

$$Pb(NO_3)_2 + H_3PO_4 \rightarrow PbHPO_4 + 2\ HNO_3\ . \tag{1}$$

Owing to the slow diffusion of lead nitrate in the gel, the reaction region advances gradually into the gel, from the gel solution interface. It forms a region of low supersaturation, i.e. a region of low nucleation efficiency where a small number of generated crystals grow in a weak competition. By using a 2.0 M concentration of lead nitrate and 1.0 M orthophosphoric acid solution, LHP crystals have been obtained. Colourless transparent LHP crystals upto $6 \times 4 \times 3\ mm^3$ in size have been grown in about 3 to 4 weeks. The crystals were harvested with the physical removal of gels. The crystals were critically examined by chemical analysis, EDAX, X-ray diffraction, scanning electron microprobe, density measurements, and thermogravimetric analysis, and were confirmed to be LHP single crystals. Figure 1 shows some typical crystals of LHP grown in the present work. This crystal belongs to the monoclinic crystal system and has lattice parameters: $a = 5.7822$ Å, $b = 6.6454$ Å, $c = 4.683$ Å, $\beta = 97.14°$, and its space group is P2/a with 2 molecules per unit cells.

## 2.2. Electrical conductivity measurements

To obtain pellet samples of LHP, the crystal grown in the laboratory were finely ground and the resulting powders were compressed into a die of diameter 1 cm and thickness 0.2–0.3 cm under a pressure of 1100 kg $cm^{-2}$ using a hand operated hydraulic press. The average particle size for sample was 150 μm and packing fraction was 0.637.

The crystal or the pellet, as the case may be, was mounted between the flat platinum electrodes in an ordinary conductinge cell. The silver paste was applied on both the experimental surfaces for firm contact. It was then enclosed in a resistance heated furnace and the temperature of the sample was monitored using a platinum rhodium thermocouple (±1 °C). The temperature of samples increased by regulating the input power. The rate of heating could be maintained linear for a reasonably large range of temperatures.

The resistance of the sample is measured by a BPL (India) Million Megohmmeter (Model EBM 115). The accuracy in the measurement of resistance in the range $10^5$ to $10^{10}\ \Omega$ is ±5%. The readings of the megohmmeter were recorded when the values became steady. The values were quite consistent and reproducible. The measured values of resistances at different temperatures were changed into electrical conductivity ($\sigma$) by knowing the area and thickness of samples. Experiments were also conducted on well faceted small crystals of LHP and dimensions of the crystals were measured with a travelling microscope. The values are reported after performing the experiments on each three samples of the same particle size and same packing fraction.

# 3. Observations and discussion

## 3.1. Studies of electrical conductivities

The Differential Scanning Calorimetry (DSC) analysis were performed on Mettler TA 2000 system with a furnace (1200 °C) using a platinum crucible and with reference

Table 3.1
Values of electrical conductivity of LHP samples at different temperatures

| Temperature ($T$) of the sample in K | electrical conductivity ($\sigma$) of the samples in $ohm^{-1}$ $cm^{-1}$ | |
|---|---|---|
| | crystal | pellet |
| 313 | $2.6371 \times 10^{-5}$ | $4.0371 \times 10^{-6}$ |
| 348 | $3.7363 \times 10^{-5}$ | $5.45 \times 10^{-6}$ |
| 408 | $5.2145 \times 10^{-5}$ | $9.8764 \times 10^{-6}$ |
| 463 | $8.3888 \times 10^{-5}$ | $4.36 \times 10^{-5}$ |
| 498 | $7.5545 \times 10^{-4}$ | $3.3069 \times 10^{-4}$ |
| 533 | $6.8125 \times 10^{-3}$ | $2.3776 \times 10^{-3}$ |
| 598 | $21.8006 \times 10^{-2}$ | $4.6136 \times 10^{-2}$ |
| 653 | $17.7510 \times 10^{-1}$ | $3.8661 \times 10^{-1}$ |
| 708 | 2.72501 | 1.0691 |
| 758 | 3.2058 | 0.8849 |
| 813 | 0.4954 | 0.2941 |
| 863 | 1.1508 | 1.0107 |
| 913 | 3.0371 | 1.3236 |
| 968 | 11.7204 | 2.81620 |

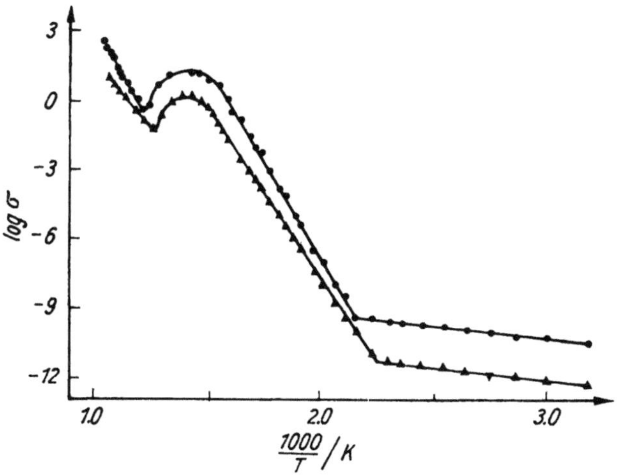


Fig. 2. The variation in the electrical conductivity with temperature for single crystal (●) and pellet (▲) samples of LHP

to $Al_2O_3$ in a nitrogen atmosphere of 1 atmospheric pressure and the scanning speed was 10 °C $min^{-1}$. The Thermo Gravimetric Analysis (TGA) were performed on DUPONT 951 analyzer with heating rate 10 °C $min^{-1}$.

The d.c. conductivity, $\sigma$ $ohm^{-1}$ $cm^{-1}$ at different temperatures for pellet as well as for single crystals of LHP are summarized in Table 3.1.

It is clearly seen from Table 3.1 that the values of $\sigma$ for crystals are always higher than for pellet samples for a given temperature over the temperature range investigated. The variation of ln $\sigma$ with the reciprocal of absolute temperature of the pellet and crystal samples is illustrated in Figure 2. The following special features should be noted:

(i) The conductivity $\sigma$ increases with increase in temperatures, till they reach the decomposition temperature.

(ii) During the decomposition temperature range 760—815 K the conductivity $\sigma$ decreases and after passing the decomposition temperature range, the $\sigma$ increases with the increase in temperature.
(iii) The variation of $\sigma$ observed in Figure 2 yields three straight lines with different slopes.

The variation in the electrical conductivity observed in Figure 2 yields different straight lines and it is in accordance with the relation:

$$\sigma = \sigma_0 e \mu \exp\left(-\frac{E_g}{kT}\right) \tag{2}$$

where
$\sigma_0$ the number of charge carrier per $cm^3$
$e$ the electronic charge
$\mu$ the mobility of the charge carrier
$E_g$ the activation energy
$k$ the Boltzmann constant
$T$ the temperature of the samples.

The linear variation of the plot is broken at 450 K temperature. So we believe that there exists three types of conduction in these materials, one at low temperature and others are at high temperatures by the way of extrinsic, intrinsic and phase change conductivity respectively in these samples.

The extrinsic conductivity is due to $(Pb)^{2+}$ and $(HPO_4)^{2-}$ ion vacancies or both. The activation energy in the extrinsic range is therefore, a measure of energy required to move the permanent defects. In the intrinsic range, where the conductivity is predominently due to movement of defects produced by thermal activation, it is associated with pure crystals. As pycnometric density and X-ray density do not differ in these samples, there may be a small concentration of Frenkel defects in the stoichiometric crystals which seems to be dominent in the electrical conduction in samples.

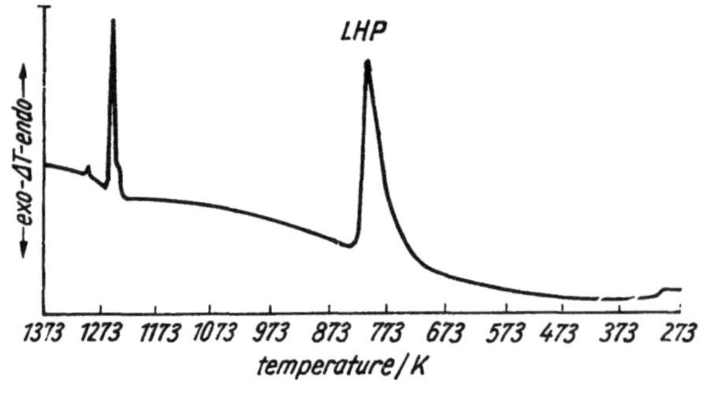


Fig. 3. DSC (differential scanning calorimetric) curve of LHP crystals

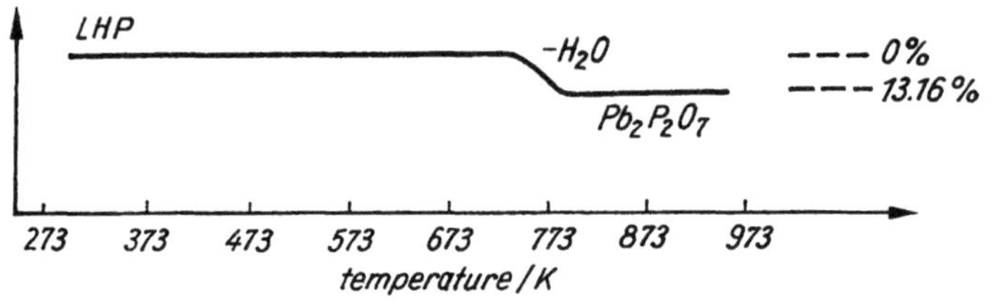


Fig. 4. TGA (Thermo Gravimetric Analysis) curve of LHP crystals

The decrease in electrical conductivity, $\sigma$ in the temperature range 760—815 K may be attributed due to the chemical decomposition. This coincides with DSC curve in Figure 3. The peak at 800 K is due to chemical decomposition and the TGA plot (Fig. 4) shows the corresponding weight loss due to decomposition is 13.16%. This may be explained due to phase change of LHP (monoclinic) into (LP) Lead Phosphate (triclinic) according to the relation:

$$2\,PbHPO_4 \rightarrow Pb_2P_2O_7 + (-H_2O)\,. \tag{3}$$

Identification of LP materials was accomplished on an X-ray powder diffractometer and the values of Bragg angles were determined using CuK$\alpha$ radiation. Using standard ASTM cards, corresponding to these values (Tab. 3.2), the material LP was identified to be triclinic. (Lattice parameters: $a = 6.9754$, $b = 12.764$, $c = 6.9627$ Å, $\alpha = 91.16°$, $\beta = 90.32°$, $\gamma = 83.22°$, Space group P $\bar{1}$).

Table 3.2

X-ray data of lead phosphate crystals obtained by heating lead hydrogen phosphate at 800 K

| Peak number | Bragg angle in degree $\theta$ | calculated values of interplanar spacing in Å, $d_{cal}$ | standard values of interplanar spacing in Å, $d_{st}$ |
|---|---|---|---|
| 1 | 6.97 | 6.344 | 6.34 |
| 2 | 9.02 | 4.911 | 4.91 |
| 3 | 9.55 | 4.644 | 4.64 |
| 4 | 11.05 | 4.02 | 4.01 |
| 5 | 11.65 | 3.815 | 3.81 |
| 6 | 12.20 | 3.646 | 3.64 |
| 7 | 12.90 | 3.451 | 3.45 |
| 8 | 13.40 | 3.324 | 3.32 |
| 9 | 14.05 | 3.172 | 3.17 |
| 10 | 14.50 | 3.077 | 3.06 |
| 11 | 14.70 | 3.036 | 3.03 |
| 12 | 15.00 | 2.976 | 2.975 |
| 13 | 15.50 | 2.880 | 2.896 |
| 14 | 16.05 | 2.786 | 2.789 |
| 15 | 16.25 | 2.753 | 2.751 |
| 16 | 17.05 | 2.627 | 2.621 |
| 17 | 17.55 | 2.555 | 2.551 |

Table 3.3

Values of activation energies for various temperature ranges for different samples

| Particulars of the sample | temperature range | activation energy (eV) |
|---|---|---|
| pellet | 348–433 K | 0.1055 |
| | 543–588 K | 0.7181 |
| | 883–953 K | 1.0771 |
| single crystals | 348–433 K | 0.2016 |
| | 543–588 K | 1.1931 |
| | 883–953 K | 1.2311 |

Crystals are more conducting than pellet samples may be attributed due to the fact that when an electric field is applied on the two opposite surfaces of the crystal, the charge carrier move freely in the crystal and reach the other electrode, while in the case of polycrystalline pellet samples, some will be trapped at the grain boundaries and give rise to interfacial polarization. The estimated values of activation energies for pellets and single crystals in different temperatures ranges are calculated and are given in Table 3.3. It is clearly seen from this table that the activation energies for crystals are higher than that of pellet samples in three temperature ranges.

### 3.2. Magnetic susceptibility measurements

The measurement of the magnetic susceptibility of LHP samples was carried outursing Faraday method (BATES). The finely powdered crystals of LHP was packed in a small container of known magnetic susceptibility. This container was hung between two powerful electromagnetic pole pieces of a Gauy balance, the sensitivity of the balance was $10^{-5}$ g. Initially, samples were weighed accurately in the absence of a magnetic field. The magnetic field was increased by increasing the current from 4 to 6 A and each time, the reduced mass of the samples were noted. The positive gram susceptibility, $\chi_g$, characteristic of a paramagnetic material was calculated using the formula

$$\chi_g = \frac{\alpha + \beta\, dw}{W} \tag{4}$$

where

$\alpha$ a constant which represents a correction factor due to the displacement of air
$\beta$ a constant for the glass container at a given current
$W$ the mass of the substance
$dw$ the change in mass when a magnetic field is applied,

i.e. $\Delta W - \delta$, $\Delta W$ is the apparent change. in the weight (in mg) of the tube containing $W$ g of the sample on the application of the field and $\delta$ is difference in weight of the empty tube on the application of the field.

The gram susceptibility of the substance was multiplied by its molecular mass to obtain molar susceptibility $\chi_M$. The susceptibility obtained at room temperature (28 °C) by this method is recorded in Table 3.4. It is evident from this table that $\chi_g$ is independent on the current.

Table 3.4
Magnetic susceptibility of LHP crystals

| Observation number | Current in A | loss in weight, $dw$ | $\alpha \times 10^{-6}$ | $\beta \times 10^{-6}$ | gram susceptibility ($\chi_g$) ($\times 10^{-9}$ e.m.u.) | molecular susceptibility ($\chi_m$) ($\times 10^{-6}$e.m.u.) |
|---|---|---|---|---|---|---|
| 1. | 4 | 0.00411 | 0.02949 | 0.7897 | 7.5165 | 2.2787 |
| 2. | 6 | 0.00651 | 0.02949 | 0.4861 | 7.5691 | 2.2947 |

## 4. Conclusions

1. LHP single crystals upto $6 \times 4 \times 3$ mm³ in size have been grown. They belong to monoclinic system and with space group P2/a.
2. The variation in d.c. conductivity of single crystal and pellet samples have been studied in the temperature range between 313–968 K.

3. The electrical conductivity of the crystalline sample is larger than that of pellet samples.
4. The activation energies for the conduction have been estimated for pellet and crystalline samples. The activation energies for crystals are higher than that of pellet samples.
5. The magnetic susceptibility of LHP was found to be $2.2862 \times 10^{-6}$ e.m.u. The positive value show that LHP is a paramagnetic materials.

The authors are thankful to Council of Scientific Industrial Research (CSIR), New Delhi for the sanction of the project, entitled "Growth and Characterization of Ferroelectric crystals". The project number and date of sanction is No. 3(564) 84 EMR II dated September 1984. One of us CCD is very much grateful to Dr. Dinesh Desai (BDS) and Mrs. Ranjan Desai (USA) for their interest and encouragement.

## References

Bates, L. F.: Modern Magnetism, London 1951
Brezina, B., Havrankova, M.: J. Cryst. Growth **34**, 248 (1976)
Desai, C. C., Rai, J. L.: Surface Technology **22**, 189 (1984)
Desai, C. C., Ramana, M. S. V.: XVII National Seminar on Crystallography, University of Madras, 22 (1985)
Dreyfus, R. W., Nowick, A. S.: Phys. Rev. **126**, 1367 (1962)
Esheby, J. D., Newey, C. W. A., Pratt, P. L.: Phil. Mag. **3**, 75 (1958)
Harris, L. B.: J. Phys. Chem. Solids **32**, 59 (1971)
Lal, H. B., Dar, N.: J. Phys. Chem. Solids **38**, 161 (1977)
Rigdon, M. A., Grace, R. E.: J. Amer. Ceramic Soc. **56**, 475 (1973)
Schulten, M. A. de: Soc. Franc. Mineral. **27**, 109 (1904)

(Received November 26, 1986)

*Author's Address*:

Professor (Dr) C. C. Desai, M. S. V. Ramana
Department of Physics, Sardar Patel University
Vallabh Vidyanagar 388120
Gujarat State, India

**Book Review**

MANFRED GRASSERBAUER, HANS JOACHIM DUDEK, MARIA F. EBEL

## Angewandte Oberflächenanalyse mit SIMS, AES und XPS

Akademieverlag Berlin, 1986, 300 Seiten, 152 Abb., 29 Tabellen, 198,— M, ISBN 3-05-500 059-5.

In dem vorliegenden Buch werden von den 3 Autoren mit Spezialkenntnissen 3 moderne bewährte Methoden der Oberflächenanalytik von Festkörpern und Werkstoffen dargeboten. Die Beiträge sind so abgefaßt, daß sie auch zur Einarbeit in das jeweilige Fachgebiet geeignet sind. Sie sind sehr eigenwillig abgefaßt, und sie sind ohne Bezug aufeinander. Es wäre zweckmäßig gewesen, wenn ein separates Kapitel über Probenpräparation und Ionenätzen vorangestellt worden wäre. Unabhängig davon wird der Werkstoff-Fachmann in diesem Buch Anregungen für die Lösung seiner Probleme finden. An einer Reihe von Beispielen mit dem Einsatz von Korrekturverfahren wird jeweils eine Vertiefung der Analysentechnik erzielt. Auch sind hinreichende Literaturangaben für ein weiterführendes Studium gegeben.

Abgesehen von einigen Druckfehlern ist die Ausstattung des Buches gut. Der Preis von 198,— M ist leider sehr hoch.

O. BRÜMMER

Cryst. Res. Technol. | 22 | 1987 | 6 | 817–825

G. Hauck

Zentralinstitut für Elektronenphysik,
Akademie der Wissenschaften der DDR, Berlin

# Surface Structure of Obliquely Deposited Layers and Liquid Crystal Alignment

The dependence of the orientational action of obliquely deposited $SiO_x$ layers on the deposition angle and the residual gas pressure has been investigated. The variation of the pressure gave rise to a modification of the general dependence of the pretilt angle of liquid crystals on the deposition angle. A model has been developed which explains the action of all deposition parameters on the orientational behaviour of liquid crystals. In a first step, some tendencies of the layer formation are stated. Thereby, in a second step, it is possible to explain the change in the orientation at variations of deposition parameters using the knowledge of these tendencies and experimentally stated informations on the relative change of layer parameters.

Es wurde die Abhängigkeit der Orientierungswirkung von schräg aufgedampften $SiO_x$-Schichten vom Bedampfungswinkel und vom Restgasdruck untersucht. Die Variation des Druckes führte zu einer Modifizierung der allgemeinen Abhängigkeit des Ankippwinkels von Flüssigkristallen vom Bedampfungswinkel. Es wurde ein Modell entwickelt, das die Wirkung aller wesentlichen Bedampfungsparameter auf das Orientierungsverhalten von Flüssigkristallen erklärt. In einem ersten Schritt werden einige Tendenzen der Schichtausbildung festgestellt. Dadurch ist es in einem zweiten Schritt möglich, die Veränderung der Orientierung bei Variation der Aufdampfparameter unter Anwendung der Kenntnis dieser Tendenzen und experimentell ermittelter Informationen über die relative Änderung von Schichtparametern zu erklären.

## 1. Introduction

It is well known that the surfaces of solid substrates exhibit an aligning influence on the molecules of liquid crystals (LC). In order to obtain a special LC alignment the substrates are often covered with layers of anorganic materials deposited at an oblique angle by evaporation in high vacuum (Janning). Layers of silicon monoxide, the most widely used evaporant, are able to induce at particular deposition angles $\theta$ (Fig. 1) a planar or a pretilted LC orientation. Electron photomicrographs (Dixon et al.; Pollack et al.) showed that these $SiO_x$ layers possess a complicated surface topography. Guyon et al. proposed a model in which the pretilt arises from the formation of a sawtooth-like structure of $SiO_x$ on which the LC alignes locally parallel to the inclined surfaces. In the bulk of the LC an average orientation is established which involves elastic deformations of the orientation pattern. These deformations depend on the shape and size of the surface structures.

Various studies have shown that thin films of different amorphous and polycrystalline materials deposited on a substrate at oblique angles exhibit anisotropic properties which are caused by the inclined columnar structure of these films. Goodman et al. varified this behaviour also in the case of $SiO_x$ and were be able to describe in this

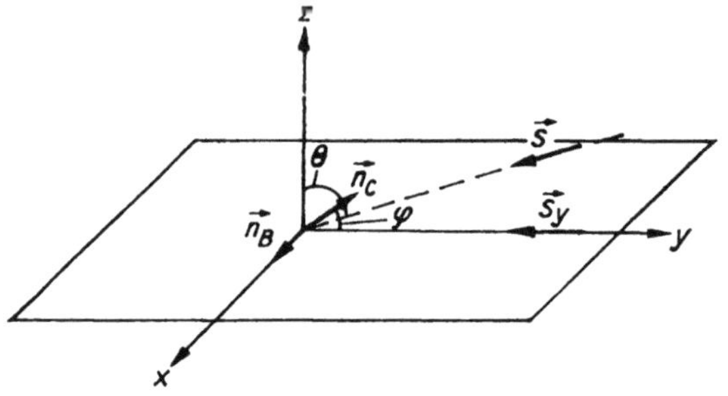


Fig. 1. Deposition direction **S** and liquid crystal alignment

way the occurrence of the planar and the pretilted orientation. However, this easy model could not explain some experimental observations done by several authors: very different pretilt angles $\varphi$ (Fig. 1) at similar LC substances, the dependence of $\varphi$ on $\theta$ and other deposition parameters, the occurrence of an intermediate $\theta$ region without a defined orientation. UCHIDA et al. removed some of these obscurities with an improved columnar model considering the dependence of the distance between the columns on $\theta$.

In general, the LC orientation on obliquely deposited layers is determined by the evaporated material and the substrate as well as by the deposition parameters ($\theta$, deposition rate $\tau$, temperatures of evaporation source $T_E$ and substrate $T_S$). There are also some evidences that the alignment is influenced by the residual gas pressure $p$ in the vacuum system (MÜLLER) and the layer thickness $d$ (UCHIDA et al.).

The purpose of this paper is to report on investigations of the influence of residual pressure on the orientation for some selected LC mixtures. Moreover, the action of the deposition parameters on the formation mechanism of the orientational layers and on the LC alignment will be discussed. In this way, the action of all parameters on $\varphi$ could be explained.

## 2. Experimental methods and results

The deposition of $SiO_x$ onto glass substrates, the LC cell conformation and the pretilt angle measurement by conoscopic methods have been described previously (HAUCK 1980; HAUCK et al.; KOMITOV et al.). A schematic illustration of the geometric relationships between deposition and alignment is shown in Figure 1.

The $SiO_x$ layers were investigated using transmission electron microscopy (TEM) of replicas and scanning electron microscopy (SEM). TEM micrographs of $SiO_x$ layers deposited at $\theta = 85°$ and different pressure are shown in Figure 2. The observed in the surface structure periods are collected in Tab. 1. For comparison, in Figure 2c a SEM micrograph of a $Al_2O_3$ layer deposition at $\theta = 85°$ (HAUCK 1982) is presented. The inclination of the columns becomes visible at a sample rotation (Fig. 2d). The inclination angle $\Gamma$ is here smaller than 60°. Therefore, from the micrograph studies we got an increase of the period of columns at decreasing $p$. In the case of $Al_2O_3$ evaporated at higher temperatures the columns are more voluminous and more separated from each other.

The LC used to investigate the action of the deposition parameters on the LC alignment was the benzoic acid phenyl esters mixture Mi 5 (VEB Spezialchemie Leipzig). The $SiO_x$ layers had a thickness of $d = 25-30$ nm. For the whole range of deposition angles $\theta = 0°-89°$ samples were prepared at a pressure before the beginning of the evaporation $p = 3 \cdot 10^{-3}$ Pa which corresponds to a pressure $p_1 = 7 \cdot 10^{-4}$ Pa

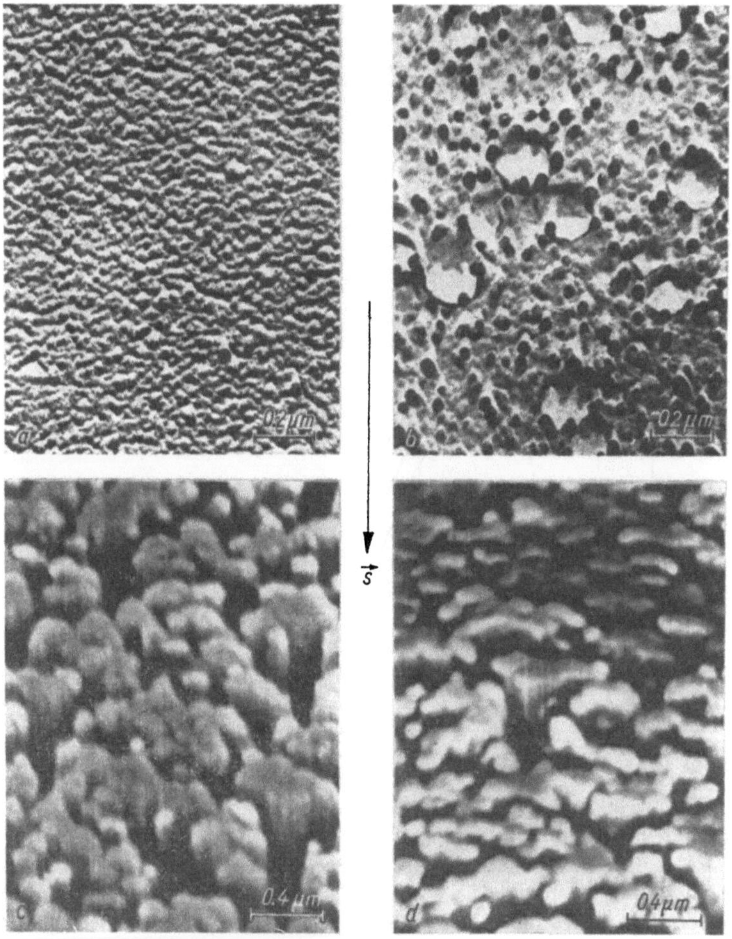

Fig. 2. Electron micrographs of obliquely deposited films with $\theta = 85°$. The arrow shows the deposition direction. a) TEM, $SiO_x$, $p = 5 \cdot 10^{-3}$ Pa, $d = 30$ nm; b) TEM, $SiO_x$, $p = 9 \cdot 10^{-4}$ Pa, $d = 30$ nm; c) SEM, $Al_2O_3$, $p = 7 \cdot 10^{-3}$ Pa, $d = 50$ nm; d) the sample shown in Figure 2c is rotated by 30° into the deposition direction

during the deposition. Moreover, in the range $\theta > 60°$ films were also deposited at $p = 7 \cdot 10^{-3}$ Pa ($p_1 = 3 \cdot 10^{-3}$ Pa) and $p = 9 \cdot 10^{-4}$ Pa ($p_1 = 3 \cdot 10^{-4}$ Pa).

By variation of $\theta$ we got three principal regions of different orientational behaviour, as known from the earlier observations of GUYON et al. and GOODMAN et al. In the range between $\theta = 0°$ and about 30° (region A) a schlieren texture indicates a planar, but degenerated orientation. In a region B ($30° > \theta > 72°$) a planar alignment in the direction $\boldsymbol{n}_B$ (Fig. 1), perpendicular to the deposition plane, is established (Fig. 3).

In an intermediate region we observed a dependence of the transition to the third orientational region (C) on $p$ (Fig. 3). At a starting pressure of $7 \cdot 10^{-3}$ Pa there is a jump between 72° and 74° to the tilted orientation $\boldsymbol{n}_C$ (Fig. 1). At $p = 3 \cdot 10^{-3}$ Pa the planar alignment remains up to 74°. Than a mixed texture composited of small areas of planar and pretilted alignment occurs. These areas are too small to measure the pretilt. Because of the statistical distribution chromatographic effects connected with the filling procedure of the cells may be excluded.

Table 1
Period between the surface structures in $Y$-direction at different residual gas pressure $p$ and layer substance

| layer substance | investigation by | $p$/Pa | period/nm |
|---|---|---|---|
| $SiO_x$ | TEM | $5 \cdot 10^{-3}$ | 35–50 |
| $SiO_x$ | SEM | $1 \cdot 10^{-3}$ | 55–75 |
| $SiO_x$ | TEM | $9 \cdot 10^{-4}$ | 60–85 |
| $Al_2O_3$ | SEM | $7 \cdot 10^{-3}$ | ~ 200 |

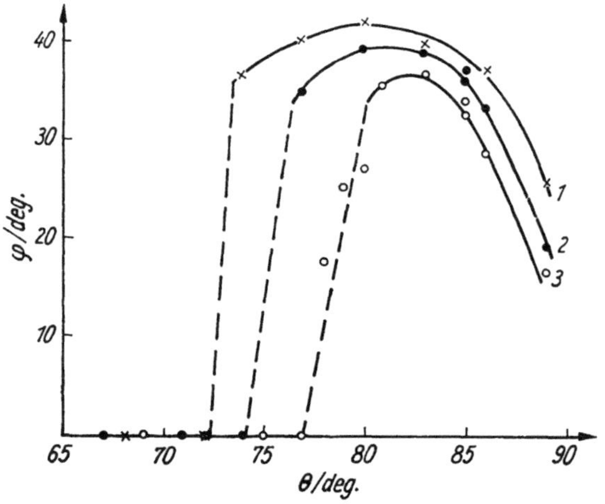


Fig. 3. Dependence of pretilt angle $\varphi$ on the deposition angle $\theta$ and residual gas pressure $p_1$, $d = 30$ nm, LC mixture Mi5 $p_1 = 3 \cdot 10^{-3}$ Pa (curve 1), $7 \cdot 10^{-4}$ Pa (2) and $3 \cdot 10^{-4}$ Pa (3)

Beginning with $\theta = 77°$ the pretilted alignment is the only one. The behaviour at $p = 9 \cdot 10^{-4}$ Pa is similar to that in the forcoming case. However, the mixed texture could be observed between $\theta = 77°$ and $81°$ and the areas are of larger size. Therefore, with a decreasing pressure the transition from the planar to the pretilted orientation is shifted to higher $\theta$ values and the mixed alignment is marked more distinctly.

In the region C the orientation is pretilted in the deposition plane, as known. At first $\varphi$ increases with the deposition angle. The following decrease is similar to the behaviour of thin deposited orientational layers described at first by UCHIDA et al. Moreover, $\varphi$ becomes smaller with decreasing $p$.

## 3. Discussion

### 3.1. Growth of obliquely deposited films (model of VAN DE WATERBEEMD)

To discuss the action of $\theta$ and $p$ on the alignment we will at first look at the formation of the surface structures which is influenced by the deposition parameters.

By GOLDSTEIN the occurrence of a film surface structure in contrary to smooth films at normal incidence is connected with self-shadowing effects.

VAN DE WATERBEEMD and VAN OOSTERHOUT formulated a theory for the growth of obliquely deposited films on glass that includes not only the effects of geometric shadowing, but also the effect of atomic mobility on the shape and development of the columnar film structure. We will briefly explain their mean considerations: In the

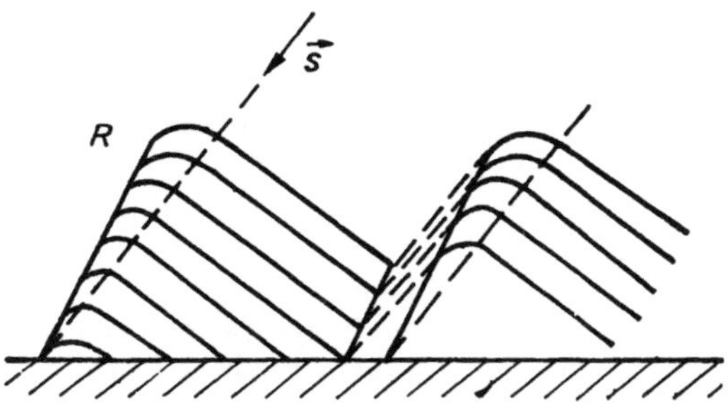


Fig. 4. Columnar growth of obliquely deposited films

initial stages of the film formation a random distribution of small nuclei is created. Their lateral growth will be realized mainly due to trapping of mobil atoms (or molecules). Because of the oblique incidence the particles have an additional mobility in the direction of the projection of vapour beam on the substrate ($\boldsymbol{S}_y$). The resulting transport along the nucleus contour causes that the point of maximum supply in front of the nucleus with respect to the vapour beam is not the point of maximum lateral growth. That later point may be defined by a critical relation between the anisotropic supply of deposited atoms along the substrate and the isotropic supply of oxygen from the background gas or by the condition of thermal equilibrium between the atoms, passing along the contour, and the nucleus. The region where the deposited atoms are trapped because the mobility becomes to low will be shifted towards the point of maximum supply be increasing $\theta$ or $p$ and by decreasing $T_E$ or $T_S$. Therefore, longshaped nucleus can be formed.

Simultaneously with the lateral growth the three-dimensional one starts. Because of the transport along the surface the growth takes place also at the shadow side (R in Fig. 4) of a nucleus, thus explaning the tilt of the shadow side towards the normal. If, however, the lateral growth causes a nucleus to reach the geometric shadow of a second one, lateral growth of the first will stop through lack of supply along the substrate. This gives rise to the columnar growth.

Investigations of obliquely deposited Ge layers confirm the great importance of the mobility of particles at the substrate and the nucleus surface (PANDYA). An increase of mobility means thereby a decrease of the ratio of number of nuclei and the area and, therefore, an increase of the diameter and the distance of the nuclei and the later formed columns.

### 3.2. Orientational layers and LC alignment at low deposition angles

The structure of obliquely deposited $SiO_x$ layers for LC alignment may be discussed as a special case of the columnar model. At $\theta = 0°$ the lateral growth is not influenced by an anisotropy, the nuclei will be circular and the layer will not have an anisotropic surface structure. The interaction between the $SiO_x$ surface and the LC molecules gives a planar alignment.

With increasing $\theta$ the lateral growth takes place particulary perpendicular to $\boldsymbol{S}$. In the beginning the width of the columnlike structures and their distance are low. The layer surface may be approximately represented as a plane with anisotropic hills: At $\theta \approx 30°$ the longshaped nuclei begin to connect together to form quasi-infinite grooves perpendicular to $\boldsymbol{S}$. If the LC molecules are aligned parallel to these rows, there are no elastic deformations in the LC near the surface and, therefore, this planar orientation in $\boldsymbol{n}_B$ is realized.

### 3.3. Growth of orientational layers at high deposition angles

At further increase of $\theta$ the particular action of each deposition parameter on the surface formation have to be taken into consideration. The corresponding, above-mentioned evidences are collected together with our results in Table 2. The vacant positions were complemented following the principles in Sect. 3.1. Because we investigate only a special system deposition material/substrate, the influence of these two parameters will be neglected.

The direction of main lateral growth defines the cross-sectional shape of the later formed columns and affects the possibility of interconnections between the nuclei in special directions. The tendencies of structural change at variations of deposition parameters are valid, naturally, only within special limits.

There are no experimental hints regarding the change in the structure of column-tops at variation of deposition conditions. Even the principle of limited mobility is not applicable to yield such informations. By reasons which will be evident in Sect. 3.4 the tops of evolved columns have probably not the shape proposed in Figure 4. We will assume, in agreement with SEM-investigations (ARMITAGE; CHENG et al.), that the cover areas of the columns are mainly parallel to the substrate plane.

From the tendencies in Table 2 it is impossible to conclude directly to the change in orientational action of obliquely deposited films. Additional informations on the special system deposited material/substrate are necessary. They concern the relative change of the single structural characteristic to each other and can be obtained experimentally. For example, the investigations of GOODMAN et al. showed that at increasing $\theta$ the distance between the $SiO_x$ columns increases more rapidly than their width. The micrographs in Sect. 2 indicate that a decrease of $p$ leads to a faster increase of the columnar width than of the distance. The same relation is valid at increasing $T_s$ (OYABU, SAITO).

Table 2

Tendencies of structural change of obliquely deposited layers at variations of deposition parameters. ↑ -increase, ↓ -decrease, (1)-VAN DE WATERBEEMD et al., (2)-PANDYA et al., (3)-GOODMAN et al., (4)-results of Sect. 2

| | ↑ $\Theta$ ↓ | ↑ $\tau$ ↓ | ↑ $p$ ↓ | ↑ $T_E$ ↓ | ↑ $T_S$ ↓ |
|---|---|---|---|---|---|
| main lateral growth at $\vec{S}$ ⟶ | ⬭ 0 (1,3) | 0 ⬭ | ⬭ 0 (1) | 0 ⬭ (1) | 0 ⬭ (1) |
| column width | ↑ ↓ (2,3) | ↓ ↑ (2) | ↓ ↑ (2,4) | ↑ ↓ (4) | ↑ ↓ (2) |
| distance between columns | ↑ ↓ (2,3) | ↓ ↑ (2) | ↓ ↑ (2,4) | ↑ ↓ (4) | ↑ ↓ (2) |
| inclination angle $\Gamma$ | ↓ ↑ (3) | ↓ ↑ (2) | ↓ ↑ | ↑ ↓ (4) | ↑ ↓ |

It is impossible to vary all deposition parameters independently from each other, e.g. the source temperature defines the evaporation rate. Therefore, the evaporation rate is often held constant and at increasing $\theta$ the deposition rate decreases. In this case, two mechanisms in Table 2 work together simultaneously. They change the characteristics of layer structure in the same direction with the only exception of $\Gamma$. But the inclination angle seems to depend not so strongly on the deposition parameters than the other column characteristics because even a great $\theta$ change affects $\Gamma$ only weakly (Goodman et al.).

### 3.4. LC alignment at high deposition angles

The LC orientation depends beside the mentioned deposition parameters also on the thickness of orientational layers (Fig. 5). At first, we will discuss the case of thick layers. At $\theta \approx 30°$ a columnar structure with chainlike interconnections is present. With increasing $\theta$ the direction of greatest nuclei extension is shifted toward $\boldsymbol{S}_y$. Hence, the interconnections perpendicular to $\boldsymbol{S}$ vanish and the orientation $\boldsymbol{n}_B$ is accompanied by an increasing deformation of the LC near the orientational layers. New interconnections parallel to $\boldsymbol{S}_y$ are impossible because the distance between the columns increases. An alignment in direction $\boldsymbol{n}_C$, also connected with an elastic deformation, will be energetically favourable. The position of the transition between the $\theta$ regions of different alignment ($\theta \sim 70°-80°$) depends due to the velocity of structural change on the deposition conditions. By this reason not one orientation is noted in Figure 5. At particular conditions the structural change takes place slowly and in a more extended $\theta$ range both orientational directions are accompanied by nearly the same elastic distortion energy. Hence, both orientations may be realized in the same cell.

In the region B the decrease of $\Gamma$ at increasing $\theta$ has no influence on the alignment. In the region C $\Gamma$ will be further decreased. The distance between the columns increases more rapidly than the width. Therefore, that part of the orientational action which aligns parallel to the substrate plane and is caused by the cover areas of the columns decreases, and $\varphi$ will be raised despite the decrease of $\Gamma$, as observed at thick layers by Uchida et al. The velocity of the $\varphi$ increase depends on the deposition parameters and due to the elasticity coefficients on the LC substance and the temperature. The pretilt angle $\varphi$ is smaller than the inclination angle $\Gamma$. If we should assume the shape of the column-tops as shown in Figure 4, the revers relationship would be valid.

At thin orientational layers the discussion of the $\varphi$ change in the region C could be carried out in the same manner up to a defined critical angle. Starting with this $\theta$ value ($\theta \approx 85°$ in Fig. 5) the influence of the substrate surface between the columns is no longer screened by the deposited structures. The part of the parallel to the $XY$-plane orienting areas will increase again and $\varphi$ decreases with further increase of $\theta$. This behaviour corresponds to the results in Sect. 2. In this way, $\varphi$ may be varied with $d$ at constant $\theta$ (Uchida et al.; Müller; Oyabu, Saito).

Uchida et al. investigated also very thin orientational layers. In this case the height of the structures is not sufficient to induce in the region C an alignment in $\boldsymbol{n}_C$ and the distance between them is too large to extend the planar alignment of the region B on the whole region C.

From the discussed mechanisms of LC orientation we can now more exactly distinguish between thick and thin $SiO_x$ layers. If the height of the columns is essentially greater than the distance, the orientational behaviour of a thick layer is present. By experimental results of Uchida et al. and Goodman et al. this is valid at $d \gtrsim 100$ nm.

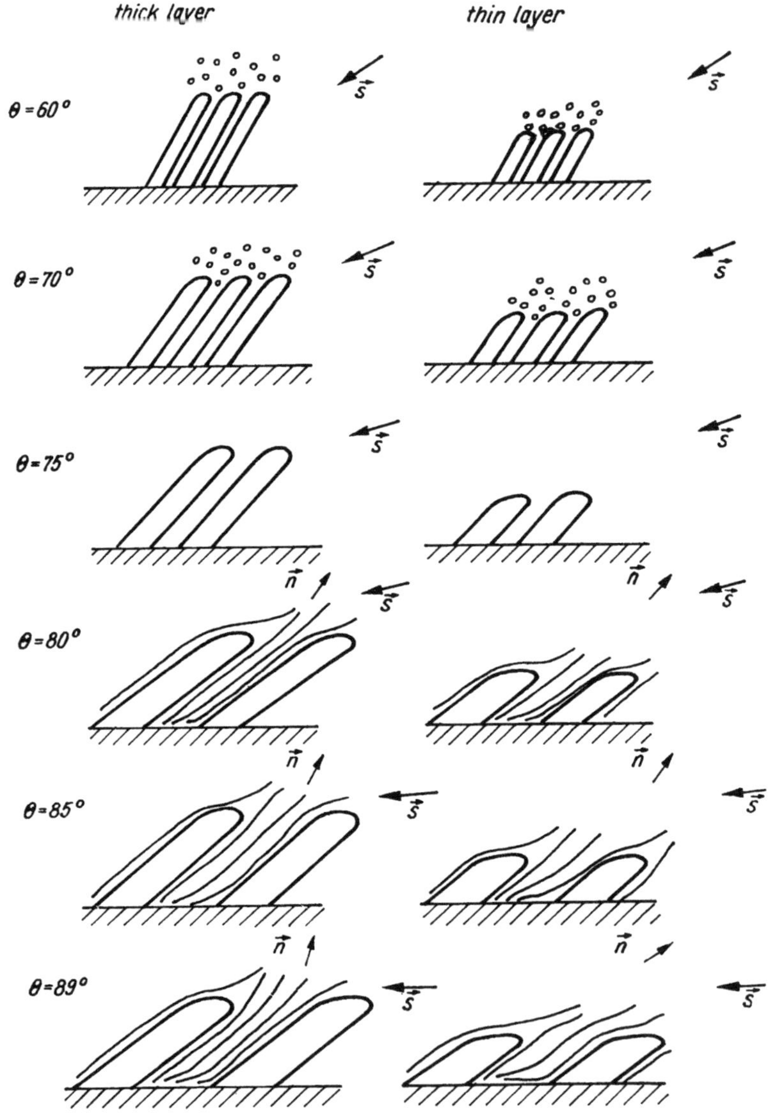


Fig. 5. Dependence of layer structure and LC alignment on deposition angle and layer thickness

At a very thin layer the height of the columns is much smaller than the distance. This case is realized at $d \lessapprox 10$ nm (UCHIDA et al.; MÜLLER).

If beside $\theta$ also other deposition parameters are changed, they modify the general $\theta$–$\varphi$-dependence. A decrease of $p$ acts due to a tendencially extention of the nuclei perpendicular to $\mathbf{S}$ (Tab. 2). The chainlike interconnections of the columns are preserved also at greater $\theta$ and the transition to the region C is shifted to greater $\theta$ values. Simultaneously the width and the distance of the columns increase. The faster change of the width gives due to the increased cover areas of the columns a decrease of $\varphi$. This behaviour corresponds to our observations (Fig. 3).

From the combination of the tendencies described in Table 2 and Figure 5 an interpretation of the different effects in the orientational behaviour of obliquely deposited

$SiO_x$ films is possible. We will discuss only two of experimentally stated facts: An increase of $T_S$ causes a decrease of $\varphi$ (OYABU, SAITO). The greater $T_S$ corresponds by Table 2 and the micrographs of OYABU and SAITO to the case of a pressure decrease and may be discussed in the same manner. An increase of $\tau$ leads also to a decrease of $\varphi$ at constant $\theta$ (KINUGAWA et al.). The tendency of extension of nuclei perpendicular to $\boldsymbol{S}$ and the decrease of columnar width and distance corresponds to the case of smaller $\theta$. Therefore, an increase of $\tau$ gives, as the decrease of $\theta$ at thick layers, a decrease of $\varphi$.

The greater size of $Al_2O_3$ surface structures (Fig. 2 c, d) might be caused beside the different system deposited material/substrate also by the higher source temperature $T_E$. Higher $T_E$ values produce by Table 2 greater formed structures.

In conclusion, the discussed tendencies of formation of orientational layers and experimentally stated informations on the relative change of characteritistc parameters of these layers enabled us to explain the dependence of LC orientation on the deposition conditions of obliquely deposited orientational films. In this way, the observed behaviour at variation of $\theta$ and $p$ could be explained.

The author wishes to thank Dr. J. Menninger for making the electron micrographs and Prof. H. D. Koswig for helpful discussions.

**References**

ARMITAGE, D.: J. Appl. Physics **51**, 2552 (1980)
CHENG, J., BOYD, G. D., STORZ, F. G.: Appl. Phys. Lett. **37**, 716 (1980)
DIXON, G. D., BRODY, T. P., HESTER, W. A.: Appl. Phys. Lett. **24**, 47 (1974)
GOLDSTEIN, B., PENSAK, L.: J. Appl. Physics **30**, 155 (1959)
GOODMAN, L. A., MCGINN, J. T., ANDERSON, C. H., DIGERONIMO, F.: IEEE Trans. Electr. Dev. **ED-24**, 795 (1977)
GUYON, E., PIERANSKI, P., BOIX, M.: Lett. Eng. Sci. **1**, 19 (1973)
HAUCK, G.: phys. stat. sol. (a) **62**, K123 (1980)
HAUCK, G.: phys. stat. sol. (a) **70**, K43 (1982)
HAUCK, G., KOMITOV, L., DERSHANZKI, A., KOSWIG, H. D.: Cryst. Res. Technol. **17**, 865 (1982)
JANNING, L. J.: Appl. Phys. Lett. **21**, 173 (1972)
KINUGAWA, K., TORIYAMA, K., TANAKA, H.: US-PS 4146656
KOMITOV, L., HAUCK, G., KOSWIG, H. D.: Cryst. Res. Technol. **19**, 253 (1984)
MÜLLER, F.: Helv. Phys. Acta **51**, 483 (1978)
OYABU, T., SAITO, S.: J. Vac. Soc. Jap. **23**, 68 (1980)
PANDYA, D. K., RASTOGI, A. C., CHOPRA, K. L.: J. Appl. Physics **46**, 2966 (1975)
POLLACK, J. M., HAAS, W. E. L., ADAMS, J. E.: J. Appl. Physics **48**, 831 (1977)
UCHIDA, T., OHGAWARA, M., WADA, M.: Jpn. J. Appl. Physics **19**, 2127 (1980)
VAN DE WATERBEEMD, J. G. W., VAN OOSTERHOUT, G. W.: Philips Res. Repts. **22**, 375 (1967)

(Received November 21, 1986)

*Author's address:*

Dr. G. HAUCK
Zentralinstitut für Elektronenphysik der Akademie der Wissenschaften der DDR
Hausvogteiplatz 5/7
Berlin
DDR-1086

**Book Review**

R. Glocker

**Materialprüfung mit Röntgenstrahlung.**
**Unter besonderer Berücksichtigung der Röntgenmetallkunde.**

Unter Mitwirkung von K. Anderko, V. Gerold, K. Kolb, E. Macherauch, S. Steeb.

Springer-Verlag Berlin-Heidelberg-New York-Tokyo. VIII + 631 Seiten mit 494 Abbildungen. Unveränderter Nachdruck der 5., erw. Aufl. 1985. Broschiert DM 78. –. ISBN 3-540-13981-8.

Die vorliegende Auflage ist der Nachdruck der Version des populären Lehrbuchs aus dem Jahre 1971. Sie war notwendig, denn nach wie vor ist der Bedarf an einer Einführung groß, die sich auf den praktischen Umgang mit Röntgenstrahlen und werkstoffkundliche Anwendungen konzentriert, dabei umfangreiche physikalische Erwägungen beiseite lassend. Und eine Überarbeitung steht noch nicht zur Verfügung.

Die erwähnte Fassung war damals von namhaften Metallforschern und Praktikern aktualisiert worden, ohne den ursprünglichen Charakter der ersten Auflage von 1927 aufzugeben. Dadurch ist der Text über weite Strecken auch heute noch als Einführung geeignet. Die inzwischen erwähnenswerten Neuentwicklungen betreffen den technischen Teil, d. h. vor allem die Strahlungsquellen (Synchrotron), die Detektoren (Ortsauflösung) und die Rechnerkopplung. Der Verlag hat sich nach Meinung des Rezensenten zu Recht nicht vom Fehlen dieser Ergänzungen abhalten lassen, den Nachdruck zu realisieren, zumal die einfachere Technik für den Anfänger durchsichtiger ist.

Als besondere Vorzüge des 'Glocker' haben zahlreiche Generationen von Materialforschern 1. die Praxisbezogenheit der Stoffauswahl und den Verzicht auf anspruchsvolle Mathematik, 2. die Behandlung der sog. Grobstrukturuntersuchung und der Spektralanalyse (die gewöhnlich von anderen Lehrbüchern ausgeklammert werden), 3. die Berücksichtigung aller Anwendungsgebiete der Röntgenstrahlinterferenzen auf dem Gebiet der Metallforschung und 4. das gute Bild- und das nützliche Tabellenmaterial (beides der Laborarbeit sehr dienlich) zu schätzen gelernt. Rund 1000 Quellenangaben erleichtern inzwischen den Zugang zur Originalliteratur umso mehr, als die Gliederung des Literaturverzeichnisses nach Sachgebieten das Lesen auch ohne Bezug auf den Text erlaubt. Alles, was der Leser dieser Einführung an mathematischem Formalismus schließlich doch zur Verfügung haben muß, enthält ein Anhang von knapp 60 Seiten. Hier findet man in sehr knapper Form alle wichtigen Themen berührt. Dabei sind allerdings Fehlinterpretationen nicht auszuschließen, wie z. B. bei der Anwendung der Beziehungen für die Netzebenenabstände, die in der angegebenen Form nur jeweils für die primitiven Zellen gelten.

Die leicht lesbare und als Broschur auch preisgünstige Auflage dürfte unter Studierenden der Werkstoffwissenschaft, Metallkunde und einzelner Spezialisierungsrichtungen der Chemie und Physik wiederum eine gute Aufnahme finden.

P. Paufler

| Cryst. Res. Technol. | 22 | 1987 | 6 | 827–833 |
|---|---|---|---|---|

K. G. SUBHADRA, K. A. HUSSAIN

Physics Department, Kakatiya University
Warangal, (A.P.), India

# Melting Temperatures of $NaClO_3$–$NaBrO_3$ Mixed Crystals

The melting points of mixed crystals of sodium chlorate and sodium bromate have been determined. The variation of melting point with composition is non-linear with negative deviations from linearity. This negative deviation from linearity is attributed to a possible increase in the vacancy concentration in mixed crystals as compared to the pure end crystals.

## 1. Introduction

Sodium chlorate ($NaClO_3$) and sodium bromate ($NaBrO_3$) are water soluble and colourless compounds. They are isomorphous, crystallizing in the cubic tetrahedral class and belong to the space group $T^4$–$P2_13$, with four molecules per unit cell. The structure lacks a centre of symmetry; consequently, these crystals are piezoelectric and optically active. $NaClO_3$ and $NaBrO_3$ are completely miscible and form solid solutions over the whole composition range.

There is very little data on the melting temperature of mixed crystals. SLAGLE and McKINSTRY (1967) have given the melting temperatures for KCl–KBr mixed crystals. CHOLOKOV et al. (1970) refer to the melting points for KCl–RbCl mixed crystals, but the actual values are not given by them. The melting points for AgCl–AgBr mixed crystals have been reported by SINGH (1970). For mixed crystals of compound substances, the data on melting is still more meagre. The melting points for the mixed crystals of $NaClO_3$–$NaBrO_3$ are not known and, as such, are determined in the present investigation.

## 2. Experimental procedure

Crystals of sodium chlorate and sodium bromate and their mixed crystals were grown from aqueous solutions. The starting materials were 99.95% pure sodium chlorate supplied by E. Merck and 99.96% pure sodium bromate supplied by Riedel-de-Haan. The materials were used in the as-supplied form without any further purification. To grow the mixed crystals, $NaClO_3$ and $NaBrO_3$ were taken in the required molar proportion, mixed and ground into fine powder. A saturated solution of this mixture was prepared and crystals allowed to grow by slow evaporation. These crystals were dried and then heated at 120 °C for 24 hours to remove any absorbed moisture. They were then ground into powder for use in melting point determination. The composition of the mixed crystals was determined by the polarographic technique. The relative concentrations of the bromate and chlorate for $NaClO_3$–$NaBrO_3$ mixed crystal system in the solid always differed from that in solution.

The melting points were determined with the help of a Boetius microscope having a micro-heating table. The sample in the form of powder is sandwiched between two cover slips and heated. The melting of the sample can be visually observed in the microscope.

The temperature of the sample can be read off from the thermometer scale, which can also be seen under the microscope along with the sample. Though the temperatures at the centre of the table may be reproducible within $\pm 0.2$ °C, since the heating is done in air, the actual uncertainty could be as large $\pm 2$ °C.

For the purpose of standardizing the procedure adopted for the determination of melting points, the melting point of sodium nitrate was determined. The melting point of sodium nitrate has been quoted in literature as 306.8 °C (Handbook of Physics and Chemistry, 1980) and 308 °C (DANIELS et al., 1970). A number of microphotographs of $NaNO_3$ crystals during the heating process were taken. Some of them are shown in Fig. (1). No change in the appearance of the minute crystals was observed upto 305 °C. Melting was observed to begin at 308 °C and was complete between 309 °C and 310 °C. The value of 309 °C was therefore taken as the melting temperature of $NaNO_3$. This value agrees within experimental error with the values reported earlier. Thus, the calibration of thermometer and the procedure adopted to locate the melting temperature seems to be satisfactory. The melting points of $NaClO_3$, $NaBrO_3$ and their mixed crystals were determined in a similar manner.

## 3. Results

i. $NaClO_3$ and $NaBrO_3$

The melting point of $NaClO_3$ as determined in the present investigation is $262 \pm 2$ °C. The transition from the crystalline state to the molten stage was clearly seen in the microphotographs taken at 260 °C and 263 °C. The melting point of $NaClO_3$ reported by different workers are given in Table 1. The present value of 262 °C is in agreement with the values reported by earlier workers.

In the case of $NaBrO_3$, the melting point quoted in standard reference book (Handbook of Physics and Chemistry, 1980) is 381 °C. $NaBrO_3$ is also known to decompose at a much lower temperature. JACH (1963) studied the thermal decomposition of $NaBrO_3$ in the temperature range 323–430 °C, and found a discontinuity in the Arrhenius plot for the rate of decomposition at about 331 °C. JACH attributed this discontinuity to the onset of melting associated with decomposition. The decomposition product NaBr forms a eutectic with $NaBrO_3$ resulting in melting below the thermodynamic melting point. No such discontinuity was observed by Jach at a temperature of 381 °C. In the present work the $NaBrO_3$ crystallites were intact till 342 °C as seen from the photographs taken at room temperature, 335 °C, 338 °C and 342 °C. A change in appearance starts at 342 °C and melting is complete at 347 °C. Some of these microphotographs are shown in Fig. (2). While our experiments enable us to observe the melting they do not preclude decomposition and its effect on melting.

Table 1
Melting points of $NaBrO_3$ and $NaClO_3$

| | melting point (°C) for | |
|---|---|---|
| | $NaBrO_3$ | $NaClO_3$ |
| JACH | $331_d$ | — |
| WATHORE and KULKARNI | $337_d$; 381 | — |
| Hand Book of Physics and Chemistry | 381 | 248–261 |
| SAWADA and SHICHIRI | — | 262 |
| present | 347 | 262 |

d denotes decomposition temperature.

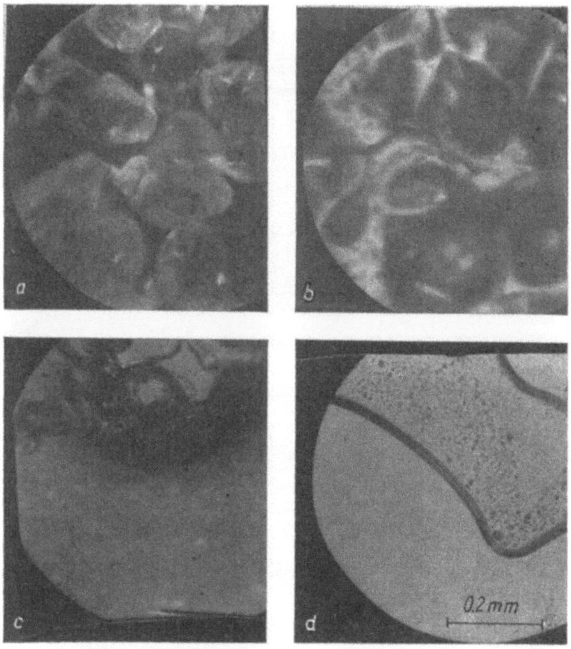


Fig. 1. Microphotographs of $NaNO_3$ at different temperatures. a) $t = 305$ °C; b) $t = 308$ °C; c) $t = 309$ °C; d) $t = 310$ °C

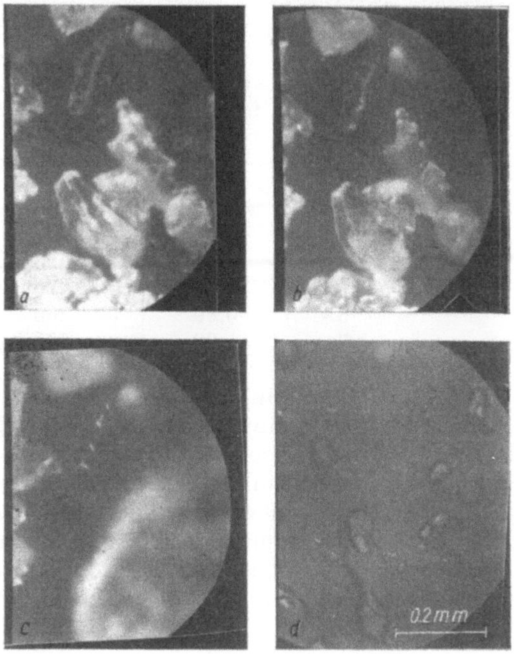


Fig. 2. Microphotographs of $NaBrO_3$ at different temperatures. a) $t = 335$ °C; b) $t = 342$ °C; c) $t = 346$ °C; d) $t = 347$ °C

ii. $NaClO_3$–$NaBrO_3$ mixed crystals

The melting points for the mixed crystals of $NaClO_3$ and $NaBrO_3$ were determined in a manner similar to that for the pure end member crystals. Typical microphotographs showing melting of mixed crystals are shown in Fig. (3) for a composition $Na(ClO_3)_{0.44}$.

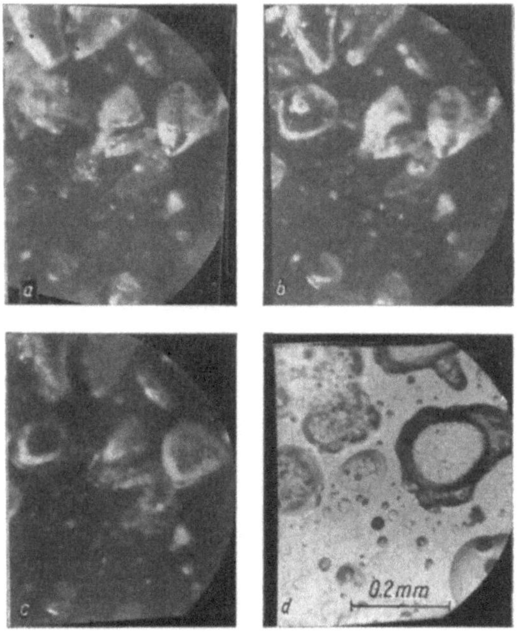


Fig. 3. Microphotographs of $Na(ClO_3)_{0.44}(BrO_3)_{0.56}$ at different temperatures. a) $t = 281$ °C; b) $t = 283$ °C; c) $t = 289$ °C; d) $t = 295$ °C

Table 2
Melting points of $Na(ClO_3)_x(BrO_3)_{1-x}$ crystals

| Sl. No. | molar fraction $x$ | melting temperature $t_m$/°C |
|---|---|---|
| 1. | 0.00 | 347 |
| 2. | 0.06 | 340 |
| 3. | 0.25 | 326 |
| 4. | 0.44 | 295 |
| 5. | 0.70 | 270 |
| 6. | 0.89 | 265 |
| 7. | 1.00 | 262 |

$\cdot(BrO_3)_{0.56}$. The melting point of the $NaClO_3$–$NaBrO_3$ mixed crystals may also be affected by the decomposition of $NaBrO_3$. It is however, not possible from our experiments to determine the effect of decomposition on the melting temperatures. The melting points of the mixed crystals are given in Table 2. The composition dependence of these melting points is slightly non-linear with negative deviations from linearity (Fig. 4). The maximum deviation from linearity is 17 °C and occurs at about the equimolal composition. The magnitude of this deviation is considerably larger than the experimental uncertainty.

## 4. Discussion

As mentioned in the preceding section, the composition dependence of melting point ($t_m$) for the $NaClO_3$–$NaBrO_3$ system is non-linear with negative deviations from linearity. This is similar to what has been observed for KCl–KBr (SLAGLE and

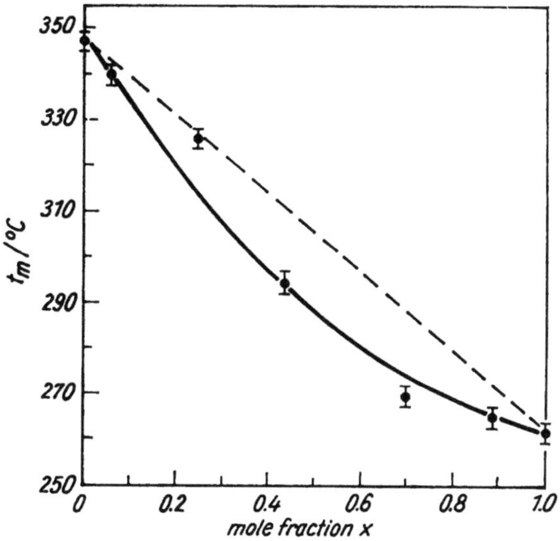


Fig. 4. Variation of melting point ($t_m$) with composition ($x$) for the $Na(ClO_3)_x(BrO_3)_{1-x}$ system Dashed line represents variation expected from additivity rule

McKinstry, 1967) and AgCl–AgBr (Singh, 1970) systems. Thus, while the magnitude of the deviation differs from system to system, the sign of the deviation from linearity seems to be a common feature for all the mixed crystal systems studied including the present one. A qualitative interpretation of this negative deviation is attempted below.

There are several approaches to the problem of melting. According to Lindemann (1910) melting takes place when the amplitude of atomic vibration becomes equal to half the interatomic spacing. Experimental data on the amplitudes of vibration at the melting point show that melting takes place at temperatures at which the amplitude of vibration has not attained the value of half the interatomic spacing. In other words, melting is taking place at a temperature lower than expected. Slagle and McKinstry estimated the melting temperatures of KCl, KBr and their mixed crystals from elastic constants making use of the Born stability criterion. The elastic constant $C_{11}$ of a cubic crystal is always greater than or equal to the elastic constant $C_{12}$. From the observed temperature variation of the quantity $(C_{11}-C_{12})$, Slagle and McKinstry estimated the melting temperature as that temperature at which $(C_{11}-C_{12})$ becomes zero. These are the predicted melting temperatures. Again, these estimated melting temperatures are much higher than those experimentally determined, which also indicates that a crystal melts at a temperature lower than the expected temperature.

Weyl (1961) suggested that no crystal melts at a temperature predicted by any lattice theory. It invariably melts at a much lower temperature. According to Weyl, the large number of vacancies at temperatures close to the melting point weakens the crystal, breaks the crystal into crystallites and clusters, and in general advances the process of melting. The lowering of the melting point from the ideal lattice value is due to the presence of vacancies.

Studies on density, ionic conductivity, dislocation densities and color centres in mixed crystals have revealed that mixed crystals have a larger number of defects than the pure end members. The density measurements of Wallace and Flinn (1953), and Wollam and Wallace (1956) on KCl–KBr and NaCl–NaBr indicate a larger concentration of vacancies in the mixed crystals than in the pure crystals. Ivankina (1958) and Makarov et al. (1960) also found that the number of vacancies in alkali halide mixed crystals far exceeds that in the pure components. The measurements of electrical conductivity of KCl–KBr by SCHULTZE (1952) indicated an increase in vacancy concentration of the mixed crystal. However, according to AMBROSE and WALLACE (1959) and ANNENKOV and GRISHUKOV (1967), the electrical conductivity of the KCl–KBr mixed crystals does not much exceed the conductivity of the pure components indicating that the vacancy concentration in the mixed crystals was not so high as the density measurements indicate.

The studies of F-centre formation in KCl–RbCl, KCl–KBr and KBr–RbBr mixed crystals by ARENDS et al. (1965) indicated a slight increase in the concentration of vacancies in the mixed crystals compared to the pure and components. In the case of mixed crystals of KCl–KBr, MELIK-GAIKAZYAN and ZAVADOVSKAYA (1960) pointed out that the variation of halfwidth of F-bands with composition can be explained by assuming a larger number of vacancies. The presence of vacancies larger in concentration than the equilibrium number in KCl–KBr mixed crystals was also confirmed by the formation of pores after prolonged high temperature annealing of this crystal (IVANKINA et al., 1960). THYAGARAJAN (1966) attributed the increased broadening of F-bands in mixed crystals of NaCl–NaBr to an increase in the density of Schottky defects compared to the pure end members.

The bulk of the evidence seems to suggest a larger concentration of vacancies in mixed crystals. This evidence taken together with WEYL's suggestion that vacancies lower the melting point accounts qualitatively for the lowering of the melting point of mixed crystals below values expected from an additivity law (negative deviations from linearity).

The reduction in melting point from the additivity value at the equimolar concentration is approximately 3 per cent in KCl–KBr and in AgCl–AgBr; and 5 per cent in $NaClO_3$–$NaBrO_3$. In the latter system, $NaBrO_3$ undergoes partial decomposition. As mentioned earlier the presence of decomposition products in the host lattice results in a lowering of the melting point. This additional factor may qualitatively account for the larger lowering of the melting point in the $NaClO_3$–$NaBrO_3$ system compared to the alkali halide mixed crystal system.

The authors wish to thank Prof. D. B. Sirdeshmukh for his encouargement.

### References

AMBROSE, T. S., WALLACE, W. E.: J. Phys. Chem. **63**, 1536 (159)
ANNENKOV, Yu. M., GRISHUKOV, V. A.: Izv. Vuz. Fiz. (USSR) **74**, 11 (1967)
ARENDS, J., DEN HARTOG, H. W., DEKKER, A. J.: phys. stat. sol. **10**, 105 (1965)
CHOLOKOV, K. S., NOVIKOV, E. N., GRISHUKOV, V. A., BOTAKI, A. A.: Izvest. Vyss. Ucheb. Zaved Fiz. **12**, 157 (1970)
DANIELS, J. M., KIRKBY, P., VIRGINIA, L. S., YUEN.: Rev. Sci. Instrum. **41**, 275 (1970)
Hand Book of Physics and Chemistry, 60th Edition, Florida 1980
IVANKINA, J. S.: Transactions of the Second All Union Conference on the Physics of dielectrics **423** (1958)
IVANKINA, J. S., ZAVADOVSKAYA, E. K., MELIK-GAIKAZYAN, I. Ya.: Fiz. Tverdogo Tel. **2**, 4 (1960)
JACH, J.: J. Phys. Chem. Solids **24**, 63 (1963)
LINDEMANN, F. A.: Z. Phys. **11**, 609 (1910)
MAKAROV, L. L., LURE, B. G., MALYSHEV, V. N.: Sov. Phys. Solid State **2**, 79 (1960)
MELIK-GAIKAZYAN, J. Ya., ZAVADOVSKAYA, E. K.: Optics and Spectroscopy **9**, 268 (1960)
SAWADA, T., SHICHIRI, J.: J. Cryst. Growth **67**, 233 (1984)
SCHULTZE, H.: Thesis, University of Göttingen (1952)
SINGH, I.: Ind. J. Technol. **8**, 52 (1970)

Slagle, O. D., McKinstry, H. A.: J. Appl. Physics **38**, 437 (1967); ibid. **38**, 446 (1967)
Thyagarajan, R.: J. Phys. Chem. Solids **27**, 218 (1966)
Wallace, W. E., Flinn, R. A.: Nature **172**, 681 (1953)
Wathore, J. N., Kulkarni, R. G.: J. Mat. Sci., **13**, 1817 (1978)
Weyl, W. A.: Advances in Chemistry Series **33**, 72 (1961)
Wollam, J. S., Wallace, W. E.: J. Phys. Chem. **60**, 1654 (1956)

(Received October 24, 1986)

*Authors' address:*

Dr. K. G. Subhadra, K. A. Hussain
Department of Physics, Kakatiya University
Warangal-506009, (A.P.), India

Cryst. Res. Technol. | 22 | 1987 | 6 | 835–844

Z. Raszewski, R. Dąbrowski, Z. Stolarzowa, J. Żmija

Military Technical Academy, Warsaw, Poland

# Dielectric Studies on Binary Mixtures Containing 4-Trans-4′-n-Hexyl-Cyclohexyl-Isothiocyanato-Benzene

We have studied binary mixtures with an induced smectic phase of 4-trans-4′-n-hexyl-cycohexyl isothiocyanatobenzene (6CHBT) with 4-n-ethyl-4′-pentylazoxybenzene (EPAB) by means of dielectric, microscopic and refractometric measurements. The effective values of components $p_\perp$ and $p_\parallel$ of the dipole moment per one molecule of the mixture have been calculated from values of $\varepsilon_\perp$ and $\varepsilon_\parallel$. As a result of our studies, we have established that the heterocomplexes formed in the mixture are so weak that they do not cause any essential changes in the distribution of molecule charges. The observed deviations from additivity of the dielectric constant of the mixture are caused solely by a non-additive behavior of the mixture density.

Wir haben ein binäres Gemisch aus 4-trans-4′-n-Hexylcyclohexylisothiocyanatbenzen (6CHBT) und 4-n-Ethyl-4′-pentylazaxybenzen (EPAB) mit smektischer Induktionsphase mittels dielektrischer, mikroskopischer und refraktrometrischer Messungen untersucht. Mit Hilfe der dielektrischen Konstanten der nematischen Phase des Gemisches $\varepsilon_\perp$ und $\varepsilon_\parallel$ wurden die effektiven Werte der Dipolmomentkomponenten $p_\perp$ und $p_\parallel$ für ein Molekül des Gemisches berechnet. Auf Grund von Untersuchungen haben wir festgestellt: Die im Gemisch entstehenden Heterokomplexe sind so schwach, daß sie keine wesentlichen Veränderungen der Ladungsverteilung der Moleküle verursachen. Die betrachteten Abweichungen von den additiven Werten der dielektrischen Konstanten des Gemisches werden nur durch nichtadditives Verhalten der Gemischdichte verursacht.

## 1. Introduction

Since the discavery of the induced smectic phase (ISP) a variety of experiments have been performed in order to understand the nature of this phase (Park et al. 1975, 1977; Yu et al.; Macmillan et al.; Griffin et al.; Chan S. Oh; Bock et al.; Szabon et al. 1982, 1983; Srikanta et al.). While studying the dielectric properties of binary mixtures made either of cyanobiphenyls and Schiff bases (Park et al.) or of cyanobiphenyls and azoxycompounds (Raszewski, Baran, Kędzierski et al.) considerable positive deviations from additivity have been observed for dielectric anisotropy.

Raszewski et al. analyzed the results of dielectric, microscopic and refractometric studies on binary mixtures containing 4-n-pentyl-4′-cyanobiphenyl (5CB) or 4-n-pentyl-4′-cyanophenyl cyclohexane (5CH). The positive deviation from linearity of the parallel component of the mean dipole moment (dipole moment per one molecule of the binary mixtures with an ISP) can be understood in terms of the increase of the number of strongly polar-weakly polar molecular pairs while the antiparallel complexes of $(5CB)_2$ or $(5CH)_2$ vanish.

However, it is very difficult in this case (when the mixtures contain cyanomesogenes the mesophases of which in the pure state are built of mixtures of single molecules and dimers of antiparallel structure (Baran, Raszewski et al.; Gray et al.; Lead-

BETTER et al.; LIPPENS et al.) to decide in what degree the changes observed in mean dipole moments of the mixture molecules are induced by formation of new hetero-complexes and in what degree by decomposition of cyanomesogene dimers. To solve the above problem we decided to carry out the study by means of dielectric and refractometric measurements on the binary mixture with an ISP formed from non-dimerizing compounds.

The mixture was as follows:

4-trans-4'-n-hexyl-cyclohexyl-isothiocyanatobenzene

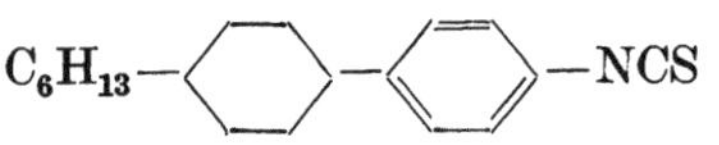


6CHBT A

with 4-n-ethyl-4'-pentylazoxybenzene

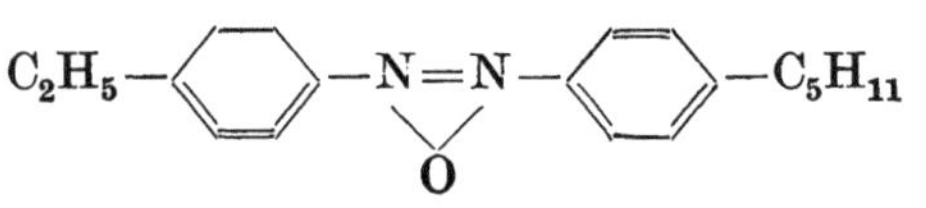


EPAB B

## 2. Experimental

The compounds of the bicomponent system investigated have been synthesized and described earlier (DĄBROWSKI et al. 1980, 1985). The mixtures were prepared by weighing and mixing thoroughly in their isotropic state.

The phase diagram was obtained by means of a REICHERT THERMOVAR microscope.

The refractive indices were measured by means of an Abbe refractometer for yellow sodium line $\lambda = 589.3$ nm, while the density of the pure compounds with a pycnometer. The temperature of the measurement prisms of the refractometer was kept constant within $\pm 0.2$ K. The surface of the measurement prism was given the nematic ordering properties by rubbing.

The dielectric constants and conductivities were measured with a TESLA bridge BM 484 at 1592 Hz, at the measuring voltage 3 V. The thicknesses of liquid crystals layers were 1 mm. To orient the samples a 1 T magnetic field was used. The temperature was stabilized within $\pm 0.2$ K. The accuracies of dielectric constants and conductivity were better than 2% and 5%, respectively.

## 3. Results

The phase diagram of the 6CHBT—EPBA mixture is plotted in Figure 1. This diagram shows a maximum in N—I thermal stability for concentration which also exhibits a maximum in the $S_A$—N curve. The phase diagram of the bicomponent system investigated is very similar in its character to that of the 5CB—EPBA mixture, but the maxima mentioned above do not lie at the equilibrium concentration, as they are moved to a composition poorer in the highly polar component.

The typical experimental curves of $\varepsilon_\perp(T)$, $\varepsilon_\parallel(T)$ and $\varepsilon_|(T)$ (dielectric permittivities in smectic, nematic and isotropic states) of a mixture of 33% 6CHBT with 67% EPAB as function of temperature are plotted in Figure 2 (part a). Principal conductivities of the same mixture are plotted in Figure 2 (part b). At all concentrations at with ISP appears both $\varepsilon_\parallel$ and $\varepsilon_\perp$ show a decrease at $T_{NS}$ while the sample is cooled. As in other cases (STRIKANTA et al.; MADHUSUDANA et al.) the anisotropy of conduc-

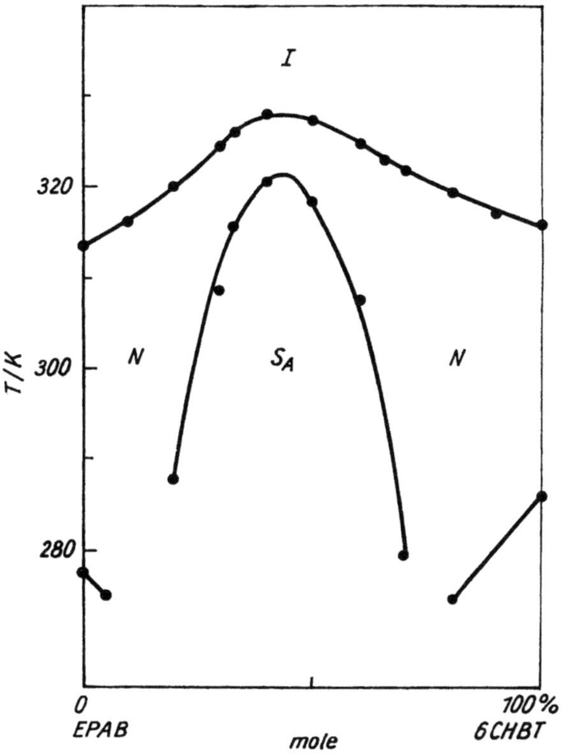


Fig. 1. The phase diagram of the 6CHBT-EPAB mixture

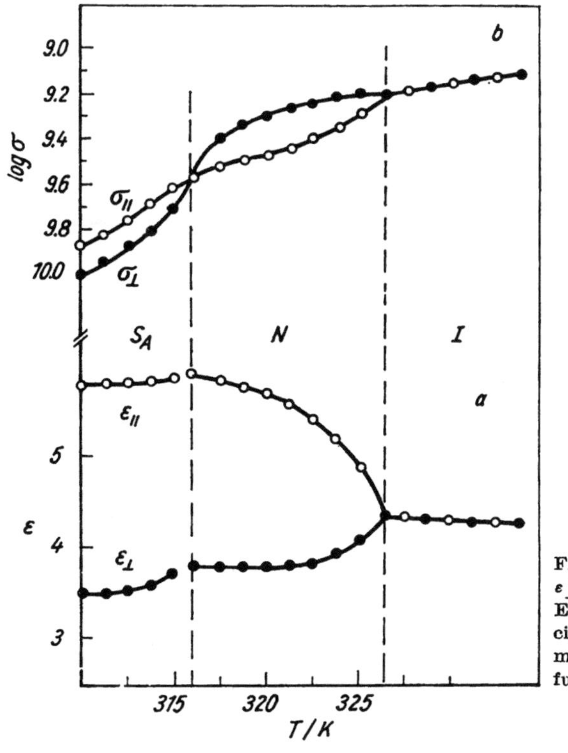


Fig. 2. Part a: Dielectric constants $\varepsilon_{\parallel}$ (○) and $\varepsilon_{\perp}$ (●) of a mixture of 33 % 6CHBT with 67 % EPAB as function of temperature; Part b: Principal conductivities $\sigma_{\parallel}$ (○) and $\sigma_{\perp}$ (●) of a mixture of 33 % 6CHBT with 67 % EPAB as function of temperature

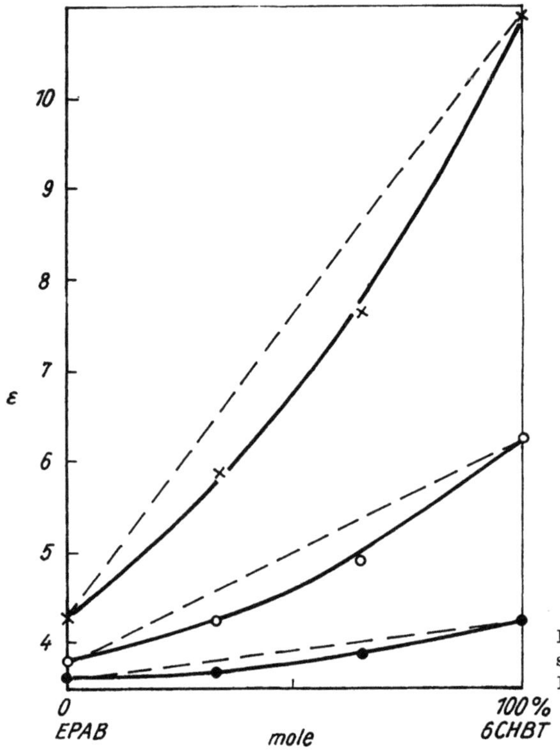


Fig. 3. Concentration changes of dielectric constants $\varepsilon_{||}$ (×), $\varepsilon_{\perp}$ (●) and $\varepsilon_{I}$ (○) of the 6CHBT-EPAB mixture

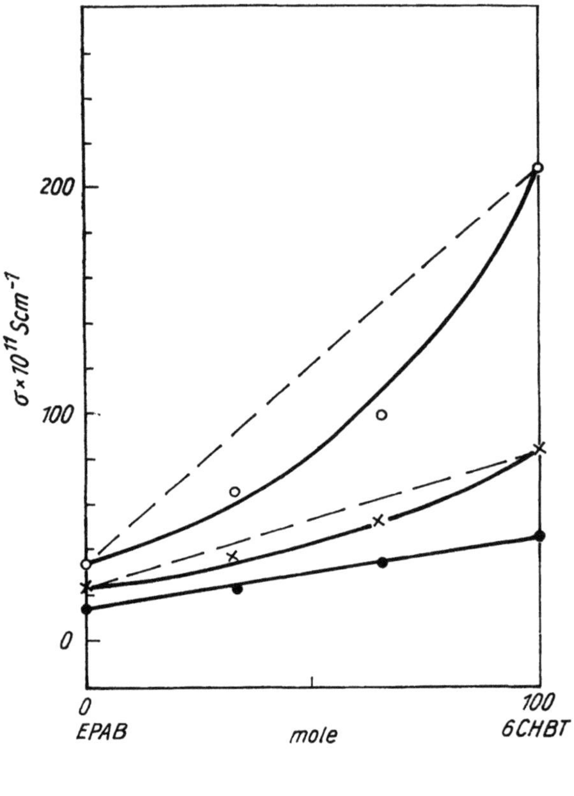


Fig. 4. Concentration changes of the conductivities $\sigma_{||}$ (×), $\sigma_{\perp}$ (●) and $\sigma_{I}$ (○) of the 6-CHBT-EPAB mixture

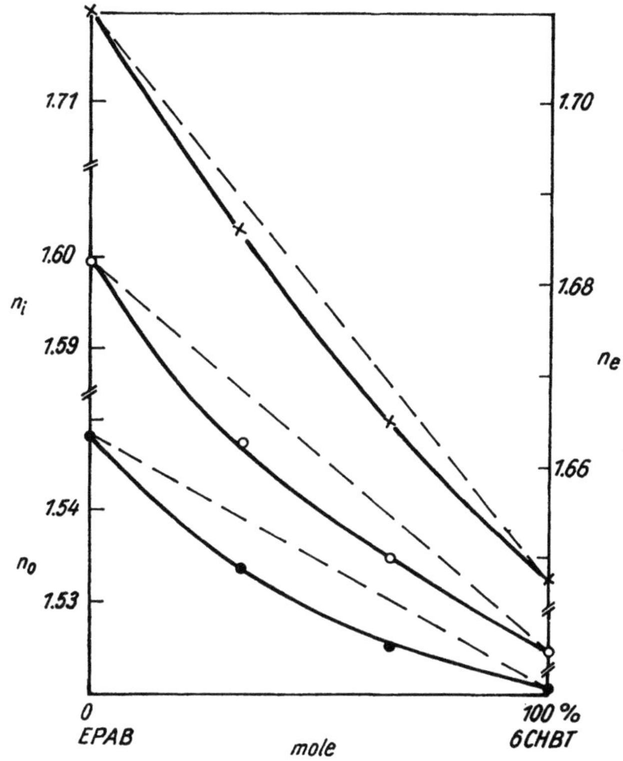


Fig. 5. Concentration changes of the refractive indices $n_e$ (×), $n_0$ (●) and $n_i$ (○) of the 6CHBT-EPAB mixture

tivity changes its sign, becoming negative a few degrees above $T_{NS}$. The parallel component $\sigma_{||}$ shows a "jump" at $T_{NA}$, while the perpendicular component $\sigma_{\perp}$ does not. This can be easily understood, because the formation of smectic layers in liquid crystals affects strongly the movement of ions in the direction perpendicular to the smectic layers. For ions moving along the direction parallel to the layers, the smectic state still seems to have the nematic structure.

The dielectric constants $\varepsilon_{||}$, $\varepsilon_{\perp}$ and $\varepsilon_I$ for the mixture investigated are plotted versus the compositions in Figure 3. Principal conductivities $\sigma_{||}$, $\sigma_{\perp}$ and $\sigma_I$, the isotropic $n_i$, ordinary $n_0$ and extraordinary $n_e$ refractive indices for the same mixture are plotted versus the compositions in Figures 4 and 5, respectively. All values in Figures 3, 4 and 5 are taken for $T = T_c \pm 10$ K.

The 6CHBT—EPAB mixture displays positive deviations from linearity for $T_{N-I}$, negative deviations for $n_i$, $n_0$, $n_e$, $\varepsilon_{\perp}$, $\varepsilon_{||}$, $\varepsilon_I$, $\sigma_{\perp}$, $\sigma_{||}$, and in practice no deviations for $\sigma_{\perp}$.

## 4. Calculations

The effective values of the components $p_{||}$, $p_{\perp}$ and $p_I$ of the molecular dipole moments of the mixture were calculated from their values of $\varepsilon_{||}$, $\varepsilon_{\perp}$ and $\varepsilon_I$ (Baran, Raszewski et al.; Baran et al. 1979, 1979) using the following equations which were obtained from W. MAIER and G. MEIER formulas

$$p_{\|}^2 = \left\{\left[\frac{3N}{\varepsilon_0(2\bar{\varepsilon}+1)}\right]^{-1} \frac{\Phi}{\bar{\varepsilon}}\left(\bar{\varepsilon} - 1 - \frac{2\,\Delta\varepsilon}{3S}\right) - \alpha_1\right\} kT\Phi\,,$$

$$p_{\perp}^2 = \left\{\left[\frac{3N}{\varepsilon_0(2\bar{\varepsilon}+1)}\right]^{-1} \frac{\Phi}{\bar{\varepsilon}}\left(\bar{\varepsilon} - 1 - \frac{\Delta\varepsilon}{3S}\right) - \alpha_t\right\} kT\Phi\,,$$

where

$$\Phi = \left[1 - \frac{3N}{\varepsilon_0(2\bar{\varepsilon}+1)} \cdot \frac{2\bar{\alpha}(\bar{\varepsilon}-1)}{9}\right];$$

$$\bar{\varepsilon} = \frac{\varepsilon_{\|} + 2\varepsilon_{\perp}}{3}; \qquad \bar{\alpha} = \frac{\alpha_1 + 2\alpha_t}{3}$$

where

$N$ numbers of molecules in 1 m³ of the investigated mixture;
$\alpha_1$ longitudinal and $\alpha_t$-transversal molecular polarizabilities.

In our calculations the molecular parameters, such as transversal $\alpha_t$ and longitudinal $\alpha_1$ polarizabilities, as well as temperature dependences of the nematic order parameter $S(T)$ for pure compounds 6CHBT and EPAB were calculated from their refractive indices $n_0(T)$, $n_e(T)$, $n_i(T)$ and densities. In these cases Neugebauer's equations combined with Haller extrapolation procedure were used. For the mixture of A and B compounds the polarizabilities $\alpha_t'$ and $\alpha_1'$ were calculated by means of the following formulas:

$$\alpha_1' = \alpha_1^A N_A + \alpha_1^B N_B\,,$$

$$\alpha_t' = \alpha_t^A N_A + \alpha_t^B N_B\,,$$

where $N_A$ and $N_B$ are the numbers of molecules of A and B compounds in 1 m³ of the mixture respectively, $N_A$ and $N_B$ were calculated from the Lorentz-Lorentz equation in the form:

$$N_A = 3\varepsilon_0 \left(\frac{n^2-1}{n^2+2}\right)\left(\alpha_A + \alpha_B \frac{M_A}{M_B} \frac{1-x_W}{x_W}\right)^{-1}$$

and

$$N_B = \frac{1-x_W}{x_W} \frac{M_A}{M_B} \cdot N_A \quad \text{so} \quad N = N_A + N_B\,,$$

where $x_W$ is the weight concentration of compound A, $M_A$ and $M_B$ are the molar weights.

$$\alpha_A = \frac{\alpha_1^A + 2\alpha_t^A}{3}; \qquad \alpha_B = \frac{\alpha_1^B + 2\alpha_t^B}{3}$$

are the mean polarizabilities, and

$$n^2 = \frac{n_e^2 + 2n_0^2}{3}$$

is the mean square of the refractive index of the mixture. The order parameters $S$ for the mixture was calculated from Neugebauer-Maier-Saupe equations as:

$$S = \frac{3}{\alpha_1' - \alpha_t'} \frac{(R\alpha - 3)\,\{[3(R\alpha + 1)]^2 - 48R\alpha\}^{1/2}}{4R}$$

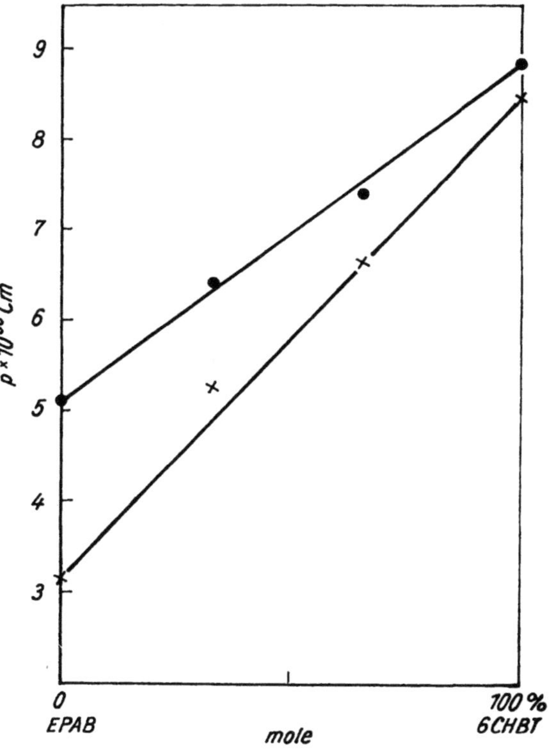


Fig. 6. Concentration dependence of the effective values of the parallel $p_{||}$ (×) components of molecular dipole moment and the total $p_c$ (●) dipole moment (calculated from nematic phase) for the 6CHBT-EPAB mixture

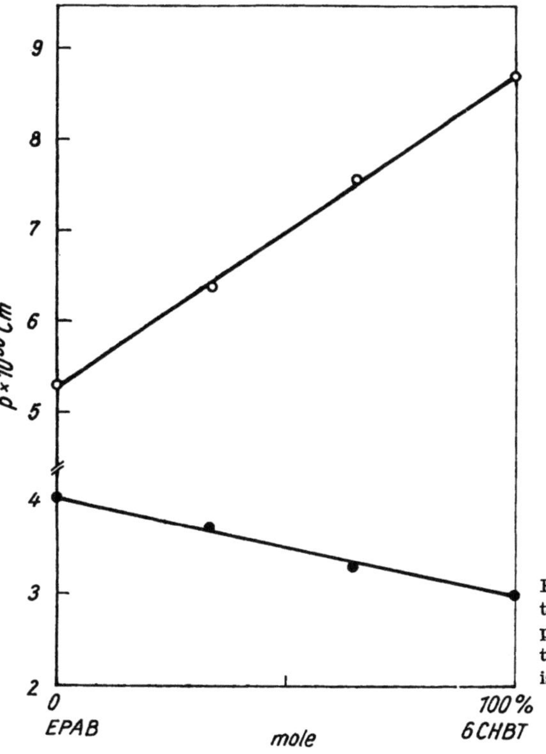


Fig. 7. Concentration dependence of the effective values of the perpendicular $p_{\perp}$ (●) components of the molecular dipole moment and the total $p_I$ (○) dipole moment (calculated from isotropic phase) for the 6CHBT-EPAB mixture

Table 1
Characteristics of the 6CHBT-EPAB mixture (all values are taken for $T' = T/ \pm 10$ K

| | EPAB | | | 6CHBT |
|---|---|---|---|---|
| $X_W/\%$ | 0 | 33.0 | 65.0 | 100 |
| $X_M/\%$ | 0 | 32.6 | 64.6 | 100 |
| $T_c$/K | 314.1 | 326.5 | 323.6 | 316.1 |
| $\varepsilon_\perp$ | 3.68 | 3.80 | 3.90 | 4.25 |
| $\varepsilon_\parallel$ | 4.30 | 5.90 | 7.55 | 10.90 |
| $\varepsilon_I$ | 3.80 | 4.25 | 4.90 | 6.25 |
| $n_0$ | 1.5490 | 1.5340 | 1.5251 | 1.5212 |
| $n_e$ | 1.7199 | 1.6970 | 1.6740 | 1.6610 |
| $n_i$ | 1.6000 | 1.5800 | 1.5670 | 1.5573 |
| $\varrho_N/g/cm^3$ | 1.037 | 1.019 | 1.008 | 1.002 |
| $\varrho_I/g/cm^3$ | 1.022 | 1.004 | 0.992 | 0.986 |
| $M$/g/mol | 296.0 | 297.8 | 299.6 | 301.5 |
| $\alpha_l'/C^2\ m^2\ J^{-1} * 10^{40}$ | 56.90 | 56.46 | 56.04 | 55.56 |
| $\alpha_t'/C^2\ m^2\ J^{-1} * 10^{40}$ | 36.86 | 37.03 | 37.20 | 37.39 |
| $S$ | 0.613 | 0.623 | 0.643 | 0.680 |
| $p_\perp * 10^{30}$/Cm | 4.05 | 3.72 | 3.31 | 3.00 |
| $p_\parallel * 10^{30}$/Cm | 3.11 | 5.27 | 6.65 | 8.41 |
| $p_C * 10^{30}$/Cm | 5.11 | 6.45 | 7.43 | 8.90 |
| $p_I * 10^{30}$/Cm | 5.26 | 6.40 | 7.54 | 8.71 |

where

$$R = \frac{N}{3\varepsilon_0}\left(\frac{n_e^2 + 2}{n_e^2 - 1} + 2\frac{n_0^2 + 2}{n_0^2 - 1}\right)$$

and

$$\alpha = \frac{\alpha_l' + 2\alpha_t'}{3}.$$

The results of calculations are summarized in Table 1 and in Figures 6 and 7.

## 6. Discussion

Park et al., Sharma et al., and Strikanta et al. suggest that the induced smectic phase appears when a charge transfer complex formation takes place between the molecules of two kinds: the highly polar component acting like the acceptor, while the other component acts like a donor. This interactions appear to lead to the formation of a layered arrangement typical of smectic phases. Wocławek et al.; Sadowska et al. suggest that 1:1 complexes that arise during ISP formation have their complexing energy lawer that the respective value for complexes usually encountered with hydrogen bonds or with charge transfer. Hence, they may be caused by van-der-Waal forces.

Baran, Raszewski et al. noticed that antiparallel correlations of molecular dipoles found for cyanomesogenes do not occur in the case of isothiocyanatomesogenes. Wocławek et al. confirmed that neither EPAB has any antiparallel correlation of dipole moments. So, assuming that charge transfer complexes or complexes with antiparallel dipole moments or other strong complexes are formed in the 6CHBT—EPAB mixtures, it is clear that same extra changes of dielectric constants of these mixtures should occur.

We observed (see Fig. 3) some negative deviation from linearity for $\varepsilon_{||}$, $\varepsilon_{\perp}$ and $\varepsilon_{I}$; but, after calculations, we can notice that the mixture investigated presents, in practice, no deviation from linearity both for $p_{||}$, $p_{\perp}$ and $p_c$ (calculated from nematic phase) and for $p_I$ (calculated from isotropic phase) (see Figures 6 and 7). The linear character of changes of the mean dipole moments of one molecule can be regarded as an evidence that neither complex with antiparallel correlation of dipole moments, nor charge transfer nor any other strong complexes are being formed. Deviations from additivity of the dielectric constant of the mixture are caused solely by a non-additive behavior of the mixture density.

The negative deviation from linearity of $\sigma_{||}$ and $\sigma_{I}$ can be interpreted as an effect of a diminished mobility of the charge carriers caused by the appearance of new heterocomplexes of course (to be in accord with the above) they must be very weak complexes which do not cause any changes in the charge distribution of the component molecules. Moreover, if the lengths of the complexes formed do not exceed those of molecules "side-to-side" complexes, their appearance would not cause any disturbances in the character of changes of $\sigma_{\perp}$ component in the function of changes in mixture concentration. This kind of behavior of $\sigma_{\perp}$ has been observed in mixture under study (see Fig. 4).

The above suggestions about a weak bonding of heterocomplex molecules in mixtures displaying an ISF are in total agreement with works by Janik et al. and by Szabon.

### References


Baran, J. W., Raszewski, Z., Dąbrowski, R., Kędzierski, J., Rutkowska, J.: Mol. Cryst. Liq. Cryst. **123**, 237 (1985)

Baran, J. W., Kędzierski, J., Raszewski, Z., Żmija, J.: Electron Technol. **12**, 107 (1979)

Baran, J. W., Kędzierski, J., Raszewski, Z., Żmija, J.: Electron Technol. **12**, 121 (1979)

Bock, M., Heppke, G., Richter, E. J., Schneider, F.: Mol. Cryst. Liq. Cryst. **45**, 221 (1978)

Chan, S. Oh.: Mol. Cryst. Liq. Cryst. **42**, 1 (1977)

Dąbrowski, R., Kenig, K., Raszewski, Z., Kędzierski, J.: Mol. Cryst. Liq. Cryst. **61**, 61 (1980)

Dąbrowski, R., Dziaduszek, J., Szczuciński, T.: Mol. Cryst. Liq. Cryst. **124**, 241 (1985)

Gray, G. W., Lydon, J. E.: Nature **252**, 221 (1974)

Griffin, A. C., Fisher, R. F., Havens, S. J.: J. Amer. Chem. Soc. **100**, 6329 (1978)

Janik, J. A., Janik, J. M., Dąbrowski, R., Otnes, K.: to be published

Leadbetter, A. J., Richardson, R. M., Colling, C. N.: J. Phys. **36**, C1-37 (1975)

Lippens, D., Parneix, J. P., Chapoton, A.: J. Phys. **38**, 1465 (1977)

Macmillan, H., Labes, M. M.: Mol. Cryst. Liq. Cryst. **56**, 7 (1979)

Madhusudana, N. V., Srikanta, B. S., Subramanya, M., Raj, Urs: Mol. Cryst. Liq. Cryst. **97**, 49 (1983)

Maier, W., Meier, G.: Z. Naturforsch. **16a**, 26 (1960)

Park, J. W., Bak, C. S., Labes, M. M.: J. Amer. Chem. Soc. **97**, 398 (1975)

Park, J. W., Labes, M. M.: Mol. Cryst. Liq. Cryst. Lett. **34**, 147 (1977)

Raszewski, Z., Baran, J. W., Kędzierski, J., Szubski, K., Żmija, J.: Conference on Crystal Growth and Liq. Cryst. Łódź, August 1986 (poster)

Raszewski, Z., Baran, J. W., Żmija, J.: 11th International Liq. Cryst. Conf., Berkeley, CA USA, July 1986 (poster)

Raszewski, Z., Baran, J. W., Żmija, J.: Conference on Crystal Growth and Liq. Cryst. Łódź, August 1986 (poster)

Sadowska, K., Żywociński, A., Stecki, J., Dąbrowski, R.: J. Phys. **43**, 1673 (1982)

Sharma, N. K., Pelzl, G., Demus, D., Weissflog, W.: Z. Phys. Chem. (Leipzig) **261**, 579 (1980)

SRIKANTA, B. S., MADHUSUDONA, N. V.: Mol. Cryst. Liq. Cryst. **99**, 203 (1983)
SZABON, J.: Mol. Cryst. Liq. Cryst. **124**, 343 (1985)
SZABON, J. DIELE, S.: Cryst. Res. Technol. **17**, 1315 (1982)
SZABON, J., PILZ, W., KOSWIG, H. D.: Cryst. Res. Technol. **18**, 519 (1983)
WACŁAWEK, W., DĄBROWSKI, R., DOMAGAŁA, A.: Mol. Cryst. Liq. Cryst. **84**, 255 (1982)
YU, L. J., LABES, M. M.: Mol. Cryst. Liq. Cryst. **54**, 1 (1979)

(Received November 10, 1986)

*Authors' addresses*

Dr. Z. RASZEWSKI, Professor R. DĄBROWSKI, Dr. Z. SOLARZOWA, Professor J. ŻMIJA
Wojskowa Akademia Techniczna
00-908 Warszawa, Poland

| Cryst. Res. Technol. | 22 | 1987 | 6 | 845–853 |
|---|---|---|---|---|

B. WEINERT, P. KIRSTEN, W. SIEGEL

**VEB Spurenmetalle Freiberg, Direktorat Entwicklung und Technologie, and Bergakademie Freiberg, Sektion Physik**

# The Influence of Crucible Moving on Growth Rate and Perfection of SSD-GaP Crystals

The influence of the crucible moving (lowering, rotation) during the SSD growth of GaP on the growth rate and the crystal quality is investigated. The crucible lowering leads to an improvement especially of the crystal quality characterized by the size of the single crystalline regions and by the homogeneity of the impurity incorporation. On the other hand, the crucible rotation yields no noticeable improvements (in the case of constant frequency) or a strong increase of the growth rate connected with a deterioration of the crystal quality (accelerated rotation).

Es wird der Einfluß der Tiegelbewegung (Absenkung, Rotation) während der SSD-Züchtung von GaP auf die Wachstumsrate und die Kristallqualität untersucht. Die Tiegelabsenkung führt insbesondere zu einer Verbesserung der Kristallqualität, die durch die Größe der einkristallinen Bereiche und durch die Homogenität des Störstelleneinbaus gekennzeichnet ist. Zusätzliche Tiegelrotation mit konstanter Drehzahl liefert keine merkliche Verbesserung. Im Falle beschleunigter Rotation erfolgt eine starke Vergrößerung der Wachstumsrate, die jedoch mit einer Verschlechterung der Kristallqualität verbunden ist.

## 1. Introduction

Advantages of the SSD (synthesis, solute diffusion) method are its manifold applicability and the good quality of the grown crystals (SUGII et al.). Essential disadvantages are the low growth rate (10–12 mg/h) and the bad reproducibility of single crystal growth (KANEKO et al. 1973; GILLESSEN, MARSHALL).

For the removal of these disadvantages up to now only few systematic investigations have been performed. KIRSTEN et al. reported about the influence of the temperature at the crystallization front and WEINERT et al. gave a calculation method for the diffusive material transport in the growth crucible.

In the present paper the influence of crucible lowering and crucible rotation on the growth rate and the crystal quality are investigated.

## 2. Growth variants

The principles of construction of a SSD growth apparatus are shown in Figure 1 (KOI; WEINERT). The red phosphorus contained at the bottom of the evacuated ampoule evaporates at the temperature $T_1$. The phosphorus vapour reacts at the surface of the liquid Ga with the latter forming a solid GaP crust. In the Ga melt saturated completely with P the solubility limit is exceeded at the crucible tip due to the lower temperature and GaP begins to crystallize. Caused by this crystallization and by the temperature gradient in the Ga melt a diffuse P transport occurs in the melt. However,

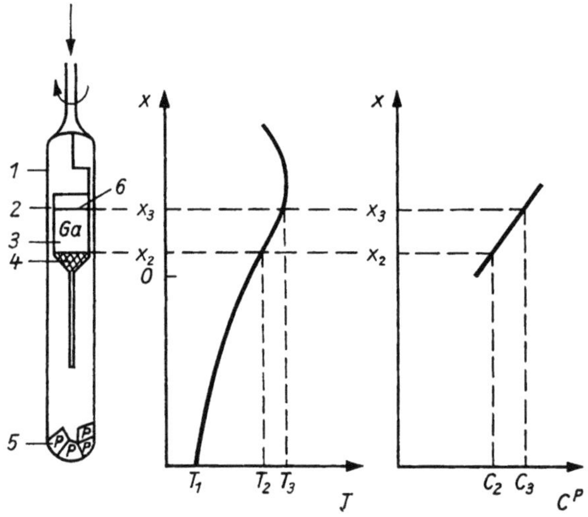


Fig. 1. SSD growth apparatus with schematic temperature and P-concentration profile in the Ga melt: 1 quartz ampoule, 2 cruciblle 3 Ga melt, 4 GaP crystal, 5 red phosphorus, 6 Ga surface with GaP crust

the transport and crystallization conditions change continuously with the growth of the GaP ingot.

By a suitable lowering of the crucible in the crystal growing furnace it is possible to counteract these changes but not to compensate these completely (RODOT et al.; MARSHALL, GILLESSEN; POIBLAUD, JACOB; KUBOTA, SUGII). The lowering rate must correspond to the growth rate. In most papers concerning the SSD method a lowering of the ampoule is used. However, the influence of this lowering on the growth rate and on the properties of the grown crystals had not been investigated in detail up to now.

An increase of the growth rate should be achieved by enforcing a defined flow in the Ga melt for the purpose of obtaining a faster P transport. By using an accelerated crucible rotation technique (ACRT) TOLKSDORF, WELZ as well as BORNMANN et al. obtained an increase of the growth rate in the case of YIG single crystal growth. GILLESSEN et al. mentioned that the application of the ACRT to the SSD growth of $A^3B^5$ materials also leads to a small increase of the growth rate. However, further particulars concerning the rotation conditions were not given.

We have realized the rotation conditions shown schematically in Figure 2. In Table 1 a survey of the different growth variants used (WEINERT) including the growth rates and crystal qualities achieved is given.

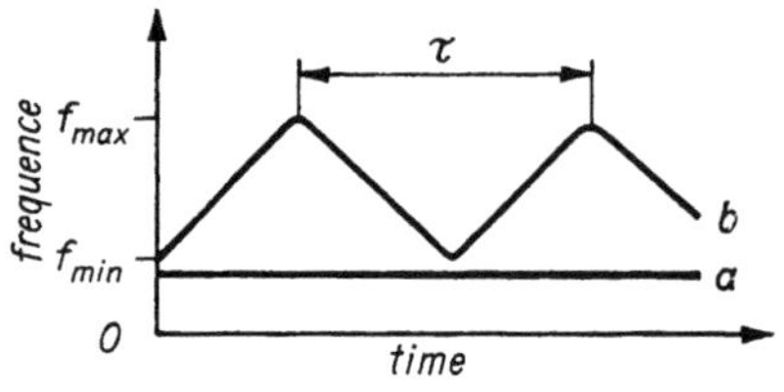


Fig. 2. Variation of the rotation frequency. a) constant frequency $f$; b) ACRT, the parameters choosed for the experiments ($f_{max} = 72\ min^{-1}$, $f_{min} = 34\ min^{-1}$ as well as $\tau = (0.1 - 0.2)$ min) were deduced from model investigations (WEINERT)

## 3. Results and discussion

### 3.1. Experiments without crucible rotation

The growth experiments with $v_{low} = 0$ yield usually polycrystalline material preferring one of the crystallographic growth directions (KOI). Figure 3a shows clearly this fact.

Table 1
Survey over all SSD experiments

| series | $T_1$/°C | $T_2$/°C | $\Delta T/\Delta x$[1]) K/cm | $v_{low}$ mm/d | rotation | $f_{min}$ min$^{-1}$ | $f_{max}$ min$^{-1}$ | $\tau$/min | $v_{gr}/v_n$[2]) | crystal perfection |
|---|---|---|---|---|---|---|---|---|---|---|
| 1 | 420 | 1000 | 25 | 0 | without rotation | $f = 0$ | | — | 1.0 | polycrystalline |
| 2 | 420 | 1000 | 25 | 1.0 | | | | — | ~ 1.2 | mostly single crystalline |
| 3 | 420 | 1000 | 25 | 1.6 | | | | — | ~ 1.4 | polycrystalline |
| 4 | 420 | 1075 | 5 | 0 | | | | — | 0.05 | polycrystalline |
| 5 | 420 | 1075 | 5 | 1.0 | | | | — | 0.25 | single crystalline with polycrystalline boundary zone |
| 6 | 420 | 1075 | 5 | 2.2 | | | | — | 0.5 | |
| 7 | 420 | 1000 | 25 | 1.0 | constant rotation | 10 | 10 | — | 1 | mostly single crystalline, crystallization front leveled, more Ga inclusions |
| 8 | 420 | 1000 | 25 | 1.0 | | 20 | 20 | — | ~ 1 | |
| 9 | 420 | 1000 | 25 | 1.0 | | 30 | 30 | — | ~ 1 | |
| 10 | 420 | 1000 | 25 | 1.0 | ACRT | 34 | 72 | 2 | ~ 1.5 | compact polycrystalline material |
| 11 | 420 | 1000 | 25 | 1.0 | | 34 | 72 | 1 | ~ 3 | porous polycrystalline material with Ga inclusions |
| 12 | 420 | 1000 | 25 | 1.0 | | 34 | 72 | 0.2 | ~ 6 | |

[1]) $\Delta T = T_3 - T_2$, $\Delta x = x_3 - x_2$, temperatures measured in the empty crucible. Concerning the temperature measurement see Kirsten et al.

[2]) $v_{gr}$ = average growth rate, $v_n$ = growth rate without lowering and rotation, $v_n = 12$ mg/h for $T_2 = 1000$ °C and $\Delta T/\Delta x = 25$ K/cm; all values refer to a crucible diameter of 13 mm

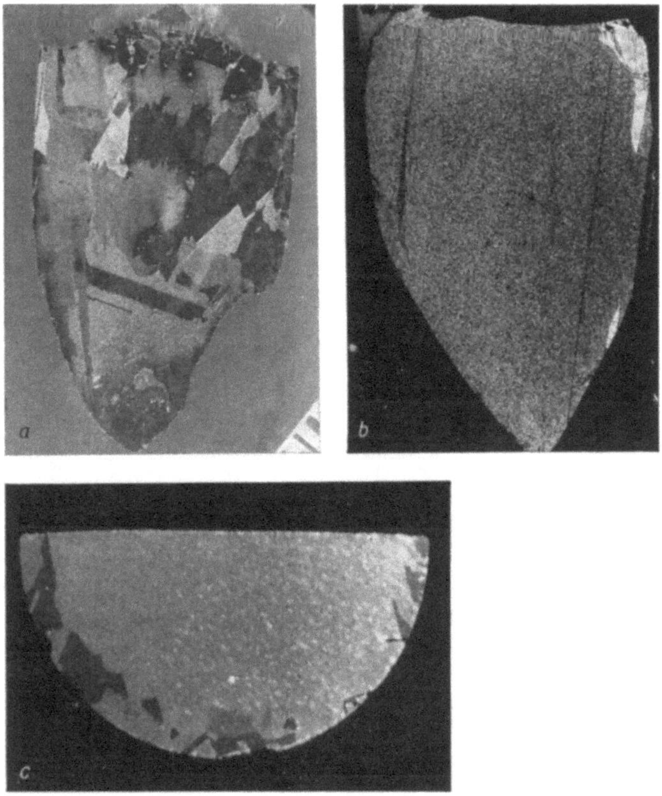


Fig. 3. Sections through GaP ingots (SSD). a) series 1, longitudinal ($T_2$ = 1000 °C), $\Delta T/\Delta x$ = = 25 K/cm, $v_{low}$ = 0); b) series 2, longitudinal ($T_2$ = 1000 °C, $\Delta T/\Delta x$ = 25 K/cm, $v_{low}$ = = 1 mm/d); c) series 6, across ($T_2$ = 1075 °C, $\Delta T/\Delta x$ = 5 K/cm, $v_{low}$ = 2.2 mm/d)

On the contrary, for all experiments with $v_{low}$ = 1 mm/d (series 2, Fig. 3b) the growth direction is near to [112].

Obviously the heat transport along the growing crystal is improved by lowering the crucible and with it the seed selection in the favourable [112] direction is promoted (RUDOLPH). If the first crystallites in the crucible tip have another orientation, then they turn over during the further growth into the [112] direction. At the crucible boundary, however, a polycrystalline region is maintained also during the further growth of the ingot (Fig. 3b). This supposes a convex phase boundary. Towards the end of the growth experiment the heat transport through the crystal worsens so that the phase boundary becomes concave. Therefore the wall crystals have better growth conditions and displace the single crystalline core.

With higher lowering rates ($v_{low}$ = 1.6 mm/d, series 3) the favourable selection conditions are not fulfilled and again polycrystalline ingots grow.

By lowering the crucible it is also possible to obtain a more stable temperature at the crystallization front. This causes a constant growth rate and a more homogeneous impurity incorporation over the length of the ingot which is demonstrated by the profiles of the carrier concentration and of the luminescence intensity along the ingot (Fig. 4 and 5, respectively). Without crucible motion the carrier concentration measured at room temperature in the crystals always p-type decreases strongly above the ingot tip (Fig. 4, curve c). With crucible lowering the hole concentration along the ingot

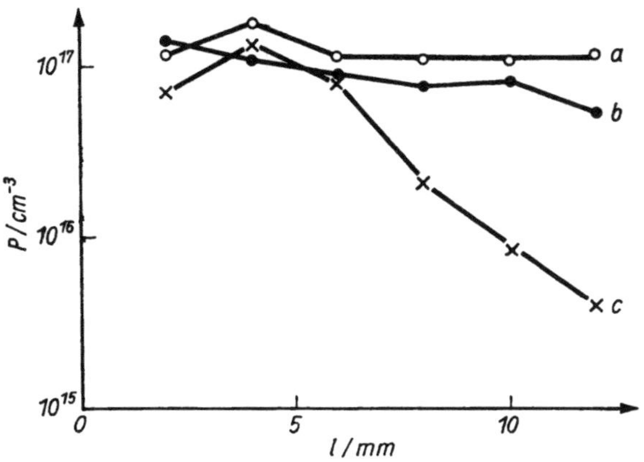


**Fig. 4.** Profiles of carrier concentration measured at room temperature along the ingots a) series 2, b) series 6, c) series 1

remains nearly constant (Fig. 4, curve a, $v_{low} = 1$ mm/d). This comparison of carrier concentrations in different ingots is reasonable because they are determined by the same impurity. This results from the analysis of temperature-dependent Hall measurements (SIEGEL et al.) which yield the concentration of the dominating acceptor carbon and that of the compensating donors (and of a medium deep acceptor present additionally). The profiles of these impurity concentrations along the ingots correspond essentially to those of the hole concentration.

The impurity concentration is also reflected by the luminescence intensity: Without crucible lowering ($v_{low} = 0$) the intensity of most lines and bands, respectively, decreases to the upper end of the ingots by about one order of magnitude (Fig. 5a, note the factors given). For $v_{low} = 1$ mm/d it changes strongly only in the ingot tip and remains nearly constant in the region in which the lowering rate corresponds approximately to the growth rate (Fig. 5b).

Growth with low temperature gradient ($\Delta T/\Delta x = 5$K/cm) and with crucible lowering yields ingots with a single crystalline core and polycrystalline boundary zone (Fig. 3c) which have a similar homogeneous impurity profile (Fig. 4, curve c and Figure 5c) as the ingots of series 2. However, the growth rate (see Tab. 1) is too low. Therefore all other experiments were performed with the parameters $T_2$, $\Delta T/\Delta x$ and $v_{low}$ of series 2.

### 3.2. Experiments with crucible rotation

A constant rotation frequency $f$ of the crucible and simultaneous lowering with the optimal lowering rate $v_{low} = 1$ mm/d yields no increase of the growth rate (series 7—9). The crystallization front being microscopically roughly for $f = 0$ is levelled at $f \neq 0$. Simultaneously the tendency for the formation of Ga inclusions increases (Fig. 6).

The application of the ARCT (series 10—12) is connected with a raising of the supply at the crystallization front. Consequently, many seeds can grow simultaneously from the crucible wall, without a selection, and only polycrystalline ingots are obtained with an increased growth rate. With decreasing cycle duration the number and size of Ga inclusions grow strongly. The crystallization front shows a characteristic scale structure (Fig. 7a). The smooth [111] faces which show striations with an average distance of 5.5 µm are favoured. These striations arise from the periodic substance transport in each ACRT cyclus. Besides stripped [112] faces (Fig. 7c) also faces with pyramidal hillocks growing in [012] direction appear (Fig. 7d). Each of these structures has a free

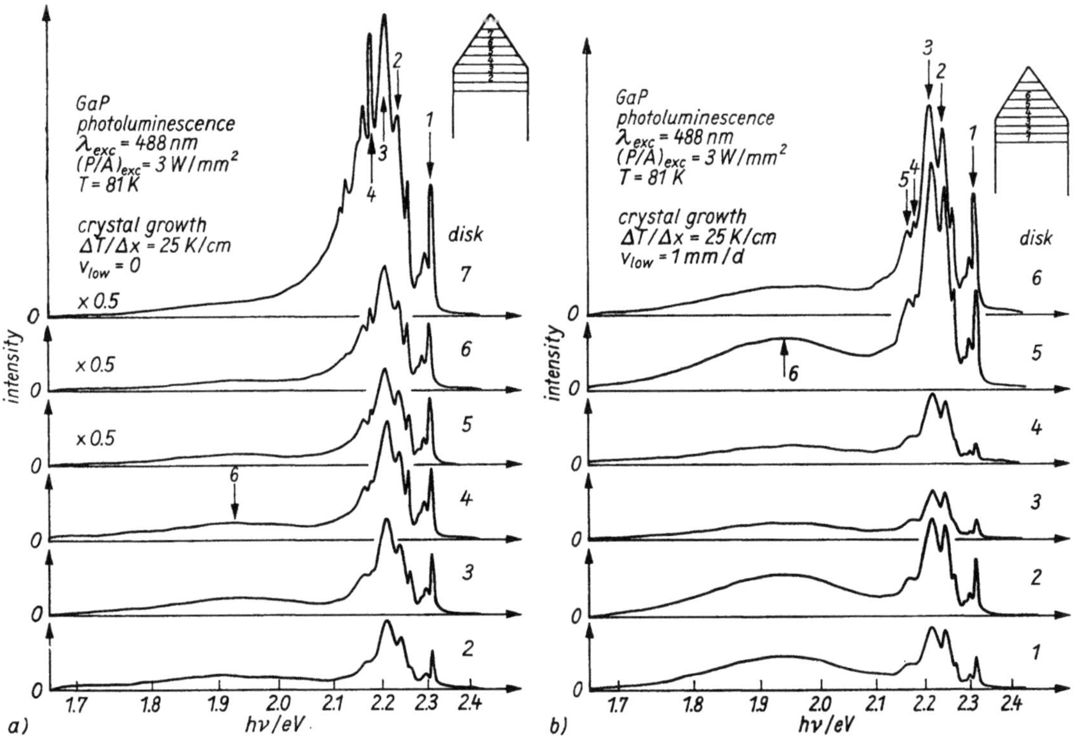


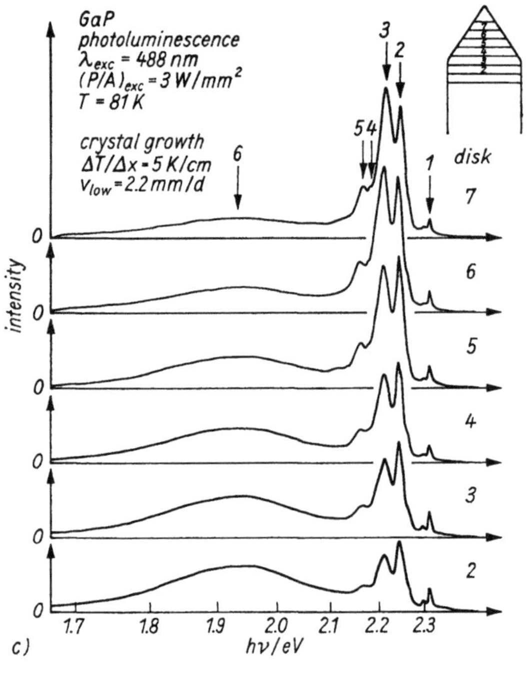


Fig. 5. Change of luminescence spectra along the ingots a) series 1, b) series 2, c) series 6. 1) A-line (exciton bound to insulated neutral N atoms); 2) free-to-bound recombination (donor S); 3) DAP recombination (donor S, acceptor C); 4) $NN_1$-line (exciton bound to two adjoining N atoms); 5) phonon replica of 3; 6) band with $E_{peak}$ = 1.93 eV and $\Delta E_{1/2}$ = 0.24 eV typically for p-GaP (SSD) (KIRSTEN et al.)

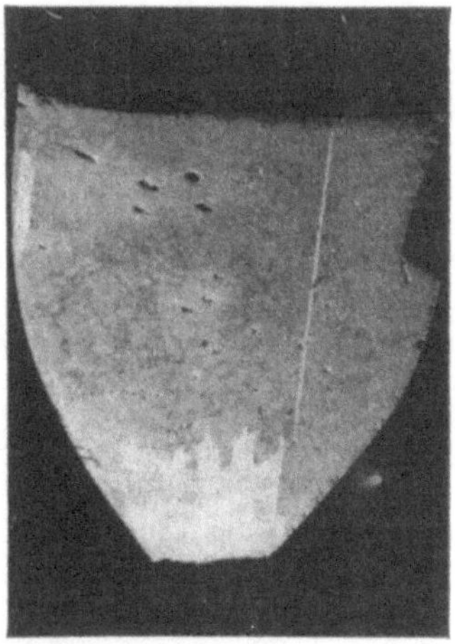

Fig. 6. Longitudinal section of a SSD-GaP ingot (series 8). ($T_2$ = 1000 °C, $\Delta T/\Delta x$ = 25 K/cm, $v_{low}$ = 1 mm/d, $f$ = 20 min$^{-1}$)

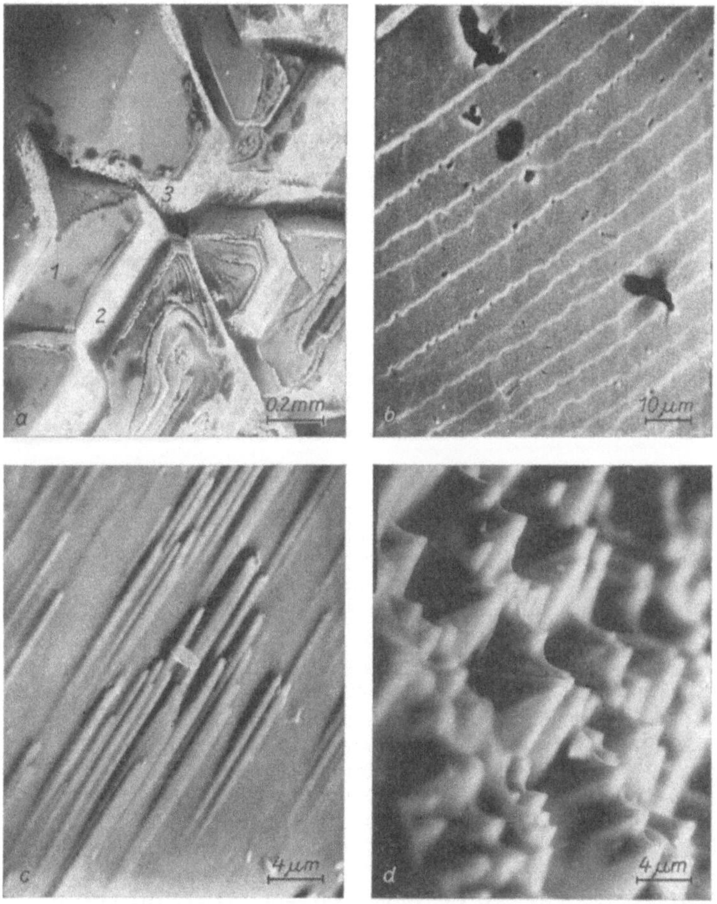


Fig. 7. REM picture of the crystallization front of a SSD-GaP ingot grown with ACRT (series 10). a) Survey photo; b) [111] face (detail of position 1); c) [112] face (detail of position 2); d) pyramidal hillocks (detail of position 3)

surface of a small single crystalline region. Because these regions cannot grow unhindered strong stresses occur. Therefore these ingots break already at low mechanical load in longish pieces which show the three characteristic faces.

Both for crucible rotation with constant frequency and for ACRT the electrical parameters of the samples taken from comparable position (ingot tip) show no significant differences compared to samples grown without crucible rotation. The luminescence of the same samples, however, is essentially lower. From this can be concluded that the incorporation of nonradiative levels increases by crucible rotation. Reasons may be an increase of the dislocation density or stressed regions (van Vechten).

## 4. Summary

By continuous crucible lowering with a rate which corresponds to the growth rate, the stability of the crystallization process of the GaP growth by the SSD method can be enhanced. The improved heat transport along the crystal causes a more favourable seed selection which leads to a preferred growth in the [112] direction with the formation of large single crystalline regions.

Due to a better constancy of the temperature at the crystallization front, moreover a homogeneous impurity incorporation over the length of the ingot is achieved.

By an additional constant rotation of the crucible the growth rate and crystal quality are not increased. The tendency to the formation of Ga inclusions increases.

By the ACRT the growth rate can be raised considerably (factor 2—6). However, the crystal quality decreases especially due to the strongly polycrystalline growth.

On the base of the results represented for the growth of single crystalline GaP by the SSD method a crucible lowering with constant rate without crucible rotation is recommended.

The authors would like to thank the group of Prof. Dr. sc. techn. Hein, as well as Dr. M. Koch and Dr. Baum (Bergakademie Freiberg, Sektion Metallurgie und Werkstofftechnik) for support. This work was performed in the "Intersektionelle Forschungsgemeinschaft Halbleiter der Bergakademie Freiberg".

## References

Bornmann, S., Glauche, E., Grönert, P. et al.: Kristall und Technik **9**, 895 (1974)
Gillessen, K., Marshall, A. J.: J. Cryst. Growth **32**, 216 (1976)
Gillessen, K., Marshall, A. J., Hesse, J.: in: Crystals: Growth, Properties and Applications, Heidelberg 1980, p. 56
Kaneko, K., Ayabe, M., Dosen, M., Morizane, K., Usui, S., Watanabe, N.: Proc. IEEE **61**, 884 (1973)
Kirsten, P., Koi, H., Siegel, W.: Cryst. Res. Technol. **20**, 1533 (1985)
Koi, H.: Diss. A, Techn. Fakultät, Bergakademie Freiberg, 1984
Kubota, E., Sugii, K.: J. Mat. Science Society of Japan **21**, 278 (1985)
Marshall, A. J., Gillessen, K.: J. Cryst. Growth **44**, 651 (1978)
Poiblaud, G., Jacob, G.: Mat. Res. Bull. **8**, 845 (1973)
Rodot, H., Hruby, A., Schneider, M.: J. Cryst. Growth **314**, 305 (1968)
Rudolph, P.: Profilzüchtung von Einkristallen, Berlin 1982
Siegel, W., Ziegler, E., Kühnel, G.: phys. stat. sol. (a) **77**, 167 (1983)
Sugii, K., Kubota, E., Iwasaki, M.: J. Cryst. Growth **46**, 289 (1979)
Tolksdorf, W., Welz, F.: J. Cryst. Growth **13/14**, 566 (1972)

van Vechten, J. A.: J. Electron. Mat. 4, 1159 (1975)
Weinert, B.: Diss. A, Techn. Fakultät, Bergakademie Freiberg 1986
Weinert, B., Buhrig, W., Hein, K.: Cryst. Res. Technol. 20, 1055 (1985)

(Received November 18, 1986)

*Authors' addresses*:

Dr. B. Weinert
VEB Spurenmetalle Freiberg
Freiberg
DDR-9200
Dr. P. Kirsten, Dr. W. Siegel
Bergakademie Freiberg
Freiberg
DDR-9200

| Cryst. Res. Technol. | 22 | 1987 | 6 | 855–859 |
|---|---|---|---|---|

Sr. Petrov, Iv. Chomakov, B. I. Bogdanov

Sofia University, Dept of Mineralogy, Sofia, and Higher-Institute of Chemical Technology Depth of Silicates, Bourgas, Bulgaria

# X-Ray Quantitative Determination of the Phases in a Batch for the Synthesis of Poly- and Monocrystal Yttriumaluminium Garnet

For the synthesis of ceramics and single crystals of yttrium aluminium garnet, a batch consisting of $Y_2O_3$ and $Al_2O_3$ is first treated thermally in order to obtain the garnet phase $Y_3Al_5O_{12}$. It is possible to make a choice of the optimum technological parameters of the batch if the amounts of all oxide phases, present during the synthesis, are well known.

A combined X-ray quantitative method is suggested for this purpose. The initial oxides and a garnet phase are determined by the method with external standard. The rest two impurity oxide phases are determined as binary system. The specific diffraction characteristics of the present compounds were taken into account. The amounts of all phases in the batch, treated thermally from 1373 to 2075 K at every 100 K are shown.

Для синтеза керамики и монокристаллов итриевоалюминиевого граната, шихта, состоящеяся из $Y_2O_3$ и $Al_2O_3$ сначала подвергается термообработки для получения гранатовой фази $Y_3Al_5O_{12}$. Выбор оптимальных технологичных параметров возможен только при определении всех окисных фаз, которые образуются во время синтеза. Для этой цели предложен комбинированный количественный рентгенографский метод. Исходные окислы и гранатовая фаза определяются методом внешнего эталона. Остальные две окисные фазы определяются как бинарная система. Учетены специфические дифракционные особености всех присуствующих соединениях. Показаны содержания всех фаз в шихту, которая была обработана термически от 1373 до 2075 К на каждые 100 К.

The use of a batch consisting of $Y_2O_3$ and $Al_2O_3$ for the synthesis of ceramics and single crystals of yttrium aluminium garnet (YAG) requires precise knowledge of its phase composition after the thermal treatment which aims to obtain the garnet phase $Y_3Al_5O_{12}$. During a high temperature synthesis the formation of other yttrium aluminium oxide compounds cannot be avoided (Toropov et al.; Bogdanov et al.) and their presence in the batch greatly affects the properties of the final product.

It is not sufficient to determine only the intervals of the formation and decomposition of the additional phases during the thermal treatment (Bogdanov et al.) in order to characterize the technological parameters of the initial batch.

This work aims to study the possibilities of applying a suitable X-ray quantitative method for determining the amount of the various phases in a batch of $Y_2O_3$ and $Al_2O_3$ treated thermally from 1373 to 2073 K.

Some studies are published in which some of the oxide phases in this system are determined quantitatively (Viechinski, Caslavsky; Morozova et al.). But there are no descriptions of the quantitative method which the authors have used.

The X-ray powder method is known to be the only reliable method for quantitative determination of the phases in a multicomponent system. However, the relationship between the amount ($C_x$) of a given phase in the system and the intensity ($I_x$) of its diffraction lines is rather indefinite (ZEVIN, ZAVIALOVA):

$$I_x = k \cdot C_x \cdot \mu^{*-1} \tag{1}$$

where $\mu^*$ is the mass absorption coefficient of the sample indicating the absorption of X-rays.

Considerable interest was provoked by the quantitative method suggested by CHUNG. It eliminates the effect of the absorption multiplier $M^*$ in Eq. 1 by using corundum ($\alpha$-$Al_2O_3$) like universal internal standard. Calculations are done by the equation:

$$C_x = \frac{I_x}{I_s} \cdot \frac{C_s}{K_x} \; ; \; K_x = \left[\frac{I_x}{I_s}\right]_{1:1} \tag{2}$$

where $I_x$ and $I_s$ are the intensities of selected diffraction lines of the phase, which should be determined and $Al_2O_3$ in the sample, respectively, while $K_x$ is a constant determined beforehand.

Although the analytical solution offered by this method is quite attractive, its application in this particular case is difficult because the amount of $Al_2O_2$ ($C_s$) which takes part in the reaction varies greatly (4—40%) and the values for $C_s$ in Eq. 2 should be found by some other method. Our repeated quantitative experiments on determining the amount of $Al_2O_3$ revealed the so called "inhibiting" effect on its diffraction lines caused by the heavy impurities. This effect is a consequence of Eq. 1 and increases proportionally to the difference in the mass absorption coefficients of the two components. Table 1 gives the differences in the values of the intensity of the same diffraction line of $Al_2O_3$ ($\mu^* = 48.34$ cm$^{-1}$) mixed with YAG ($\mu^* = 109.87$ cm$^{-1}$) and $Y_2O_3$ ($\mu^* = 155.57$ cm$^{-1}$). As it can be seen, the experimental error in the quantitative X-ray determination reaches up to 10%, which makes it inexpediently to apply CHUNG's method in this case. Our opinion, however, is that in spite of some advantages of this method, further investigations are needed on its applicability to similar systems, eventually by introducing corrective coefficients in Eq. 2.

That is why the approach used in this study is the one employing an external standard. The amount of each phase under study was obtained by comparing the intensities of its lines in the sample ($I_x$) and in the pure state ($I_0$) with taking into consideration the absorption coefficient $K_m$ by the equation:

$$C_x = \frac{I_x}{I_0} \cdot K_m \; ; \qquad K_m = \frac{\mu^*}{\mu_x^*} \tag{3}$$

where $\mu^*$ and $\mu_x^*$ are the mass absorption coefficients of the sample and the phase, which is being determined, in the pure state. This method, though it is more labour-

Table 1
Differences in the intensity of $Al_2O_3$ in different "media"

| content $Al_2O_3$/% | intensity of $Al_2O_3$ peak imps/s in mixtures | | $\Delta I$ imp/s | $\Delta C$/% |
|---|---|---|---|---|
| | with $Y_2O_3$ | with YAG | | |
| 10 | 1564 | 2425 | 861 | 2.7 |
| 20 | 3035 | 4662 | 1627 | 4.6 |
| 30 | 5850 | 6237 | 387 | 1.1 |
| 50 | 9671 | 13159 | 3488 | 9.9 |
| 70 | 17328 | 18437 | 1109 | 3.2 |
| 80 | 22250 | 23556 | 1306 | 3.8 |

consuming, is characterized by high accuracy and reproducibility of the results. Practice has shown that the effect of $K_m$ on the reliability of the analyses increases considerably in the presence of absorption differences between the components of the system. This is exactly the case under study. The preliminary qualitative X-ray phase analysis (BOGDANOV et al), proves the presence of five crystal phases. In various quantities they are present in all stages of the heat treatment. Part of the diffraction pattern of a sample from a batch treated thermally at 1673 K is given in Figure 1. It also shows the reflections by which the amount of each phase was determined. The chemical composition and the mass absorption coefficients of the compounds, taking part in the reaction, are presented in Table 2.

It is clear that preliminary quantitative analyses of artificial mixtures are needed to establish the exact deviations from the rectilinear curve in Eq. 3 which are due to the effect of $K_m$. For that purpose, mixtures obtained by consecutive mixing of different amounts of $AlO_3$ with YAG (3:5), $Y_2O_3$ and a "medium" consisting of equal amounts of YAG and $Y_2O_3$, were repeatedly prepared and X-rayed. The standard plot obtained for determining $Al_2O_3$ and $Y_2O_3$ is presented in Figure 2. It shows the expected (ZEVIN, ZAVIALOVA) deviation from the rectilinear curve according to Eq. 3, which is explicable for cases of determining mixtures of light substances in a heavier "medium" ($Al_2O_3$) and vice versa ($Y_2O_3$).

If the chemistry and the amount of the compounds in the system are known, $K_m$ will have a definite value and Eq. 3 will allow a more accurate analytical solution. In the system under study this is not possible because of the indefiniteness of $\mu^*$ for the vari-

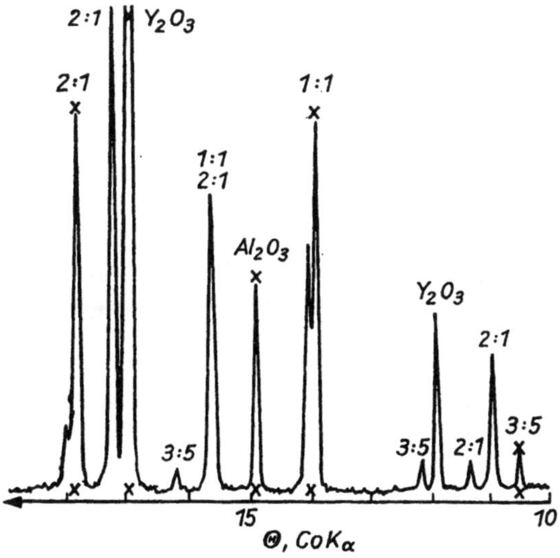


Fig. 1. Part of the diffraction pattern of the investigated system

Table 2
Chemical composition and $\mu^*$ values for the phases in the system shown on Figure 1 (for Co–$K_\alpha$ radiation)

| No. | phase | $\mu^*/cm^{-1}$ |
|---|---|---|
| 1. | $Al_2O_3$ | 48.34 |
| 2. | $Y_2O_3$ | 155.57 |
| 3. | $YAlO_3$ (1:1) | 122.50 |
| 4. | $Y_4Al_2O_9$ (2:1) | 136.17 |
| 5. | $Y_3Al_5O_{12}$ (3:5) | 109.87 |

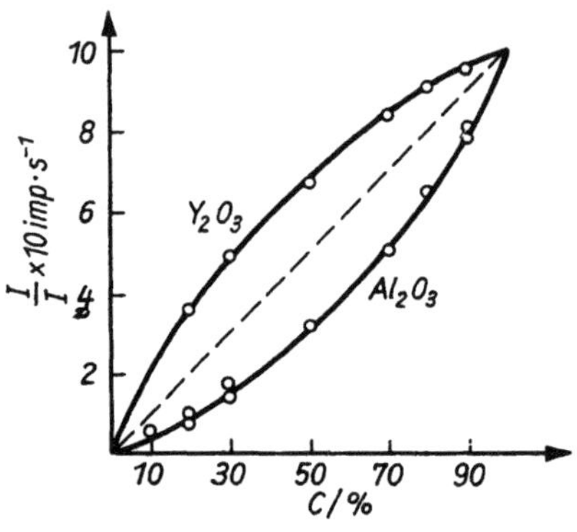

Fig. 2. Standard plots for determining $Al_2O_3$ and $Y_2O_3$

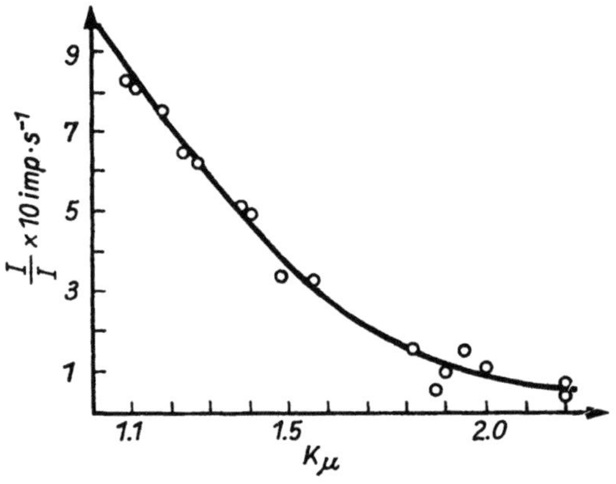

Fig. 3. Experimental plot for determining $K_\mu$, obtained by analyzing mixtures with different composition

ous samples. The analysis of a great amount of artificial mixtures made it possible to study graphically the dependence of $K$ on the intensity ratio $I_x/I_0$. In the case of $Al_2O_3$ this dependence for mixtures in different media is shown in Figure 3. The plot obtained allows the immediate determination of the $K_m$ value for each sample and to make a rapid analytical solution of Eq. 3.

By means of the X-ray method described above, the amounts of YAG, $Y_2O_3$ and $Al_2O_3$ in different series of technological samples were determined repeatedly at definite stages of the thermal treatment of the batch.

It was essentially more difficult to calculate the amount of the two impurity stages, $YAlO_3$ and $Y_4Al_2O_9$. Since these compounds are intermediate reaction products it is neither possible to synthesize them in the pure state nor to separate them by conventional means as standards by the above method. In order to determine their amount, another method was employed. It is based on the equation used for determining binary systems (ZEVIN, ZAVIALOVA):

$$\frac{C_x}{C_y} = K_m \cdot \frac{I_x}{I_y}\,; \qquad C_x + C_y = 100 - \sum_{i=1}^{3} C_i \tag{4}$$

where $C_i$ is the sum of the three phases determined by the method with an external standard.

The suggested quantitative X-ray method was tested on technological samples from a batch treated thermally to obtain ceramics and single crystals. The amounts of all phases of the system in the samples which were thermally treated at every

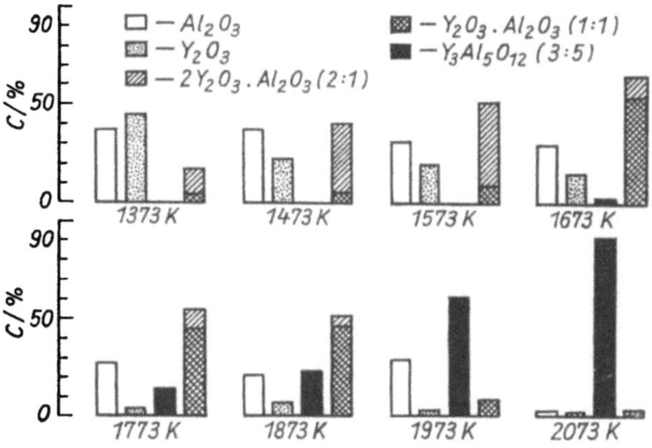


Fig. 4. Amounts of the phases at different stages of the thermal treatment

100 K, in the interval mentioned above, were determined. The results obtained are given in Figure 4. Besides the temperatures of the formation and decomposition of the intermediate compounds, Figure 4 shows also the change of the amount of the initial oxides ($Al_2O_3$ and $Y_2O_3$) at definite stages of the synthesis. It is particularly important that this quantitative analysis makes it possible to determine precisely the phase composition of the batch at the different stages of the thermal treatment. This allows purposeful thermal preparation of the batch for the synthesis of the ceramical or single crystal products.

## References


Bogdanov, B. I., Chomakov, I., Iliev, I.: Silikat Techn. (in print)

Chung, F. H.: Adv. in X-ray Anal. **17**, 106 (1973)

Morozova, L. P., Lukin, E. S., Ephimovskaia, T. V., Smolia, A. V.: Steklo i ceramika **8**, 28 (1978)

Toropov, N. A., Bondar, I. A., Galahov, F. A., Nikosian, H. S., Vinogradova, N. V.: Izv. AN SSSR, ser. CHIM. **7**, 1158 (1962)

Viechincki, D., Caslavsky, J.: J. Amer. Ceram. Soc. Bull. **58**, (8) 790 (1979)

Zevin, L. S., Zavialova, L. L.: Kolitchestveni Rentgenographitcheski Analis, Moskva, 1973


(Received October 28, 1986)

*Authors' address*:

c/o Dr. Sr. Petrov
Sofia University, Dept of Mineralogy and Crystallography
Boulvd Russki, 15
1000 Sofia, Bulgaria

Cryst. Res. Technol. | 22 | 1987 | 6 | 861–866

R. DĄBROWSKI, J. JADŻYN, J. PRZEDMOJSKI, J. BARAN

Military Technical Academy, Warsaw, Poland, and
Institute of Molecular Physics, Polish Academy of Sciences, Poznań, Poland, and
Institute of Physics, Warsaw Technical University, Warsaw, Poland

# The Nematic Reentrant Phase in Binary Mixtures Composed of a 4-(trans-4-Alkylcyclohexyl)-1-(2-Cyanoethenyl)-Benzene and 6OCB

It is shown that the smectic A phase in the 4-(trans-4-n-alkylcyclohexyl)-1-(2-cyanoethenyl)-benzene EPCH homologous series has a partially dimolecular structure of $A_d$ type. The EPCH compounds show a virtual $S_{Ad} \rightarrow N_{re}$ transition revealed in mixtures of 6OCB with members of the homologous series with five or more carbon atoms in the alkyl chain. The temperature dependences of electric conductivity and electric permittivity of the 8EPCH-6OCB mixture have been tested in the three successive phases, $N_{re}$, $S_{Ad}$ and N.

Es wurde bewiesen, daß die smektische Phase A in der homologen Reihe von 4-(trans-4-n-Alkylcyclohexyl)-1-(2-cyanethenyl)benzen EPCH teilweise bimolekulare Struktur des Typs $A_d$ hat. EPCH-Verbindungen weisen den virtuellen Übergang $S_{Ad} \rightarrow N_{re}$ auf, der im Gemisch mit 6OCB für die Glieder der homologen Reihe mit 5 und mehr Kohlenstoffatomen in der Alkylkette entdeckt wurde. Die elektrische Leitfähigkeit und die Dielektrizitätskonstanten wurden in Abhängigkeit von der Temperatur für das Gemisch 8EPCH-6OCB in den drei aufeinander folgenden Phasen $N_{re}$, $S_{Ad}$, N untersucht.

## 1. Introduction

In many binary mixtures made up of binuclear compounds with CN terminal group the appearance of the reentrant nematic phase was observed (GÖBL-WUNSCH et al.). Such a property is revealed by compounds with an alkyl chain of 7 or more carbon atoms. A similar regularity was observed for threenuclear compounds that reveal a monotropic or enantiotropic reentrant nematic phase in the pure state (NGUYEN HUU TINH). The reentrant nematic phase in mixtures and pure compounds was always observed in the first and successive members of the homologous series with the smectic $A_d$ phase. Hence the question arises whether the occurrence of the reentrant nematic phase is related with the smectic $A_d$ phase and the length of the alkyl chain, or else only with the presence of the smectic $A_d$ phase. May it be therefore observed also in compounds with alkyl substituents with less than 7 carbon atoms if they reveal the smectic $A_d$ phase.

In order to find an answer to this question we investigated the structure of the smectic phase A of compounds belonging to the 4-(trans-4′-n-alkylcyclohexyl)-1-(2-cyanoethenyl)-benzene (EPCH) homologous series (formula 1) and the properties of their binary mixtures with 6OCB.

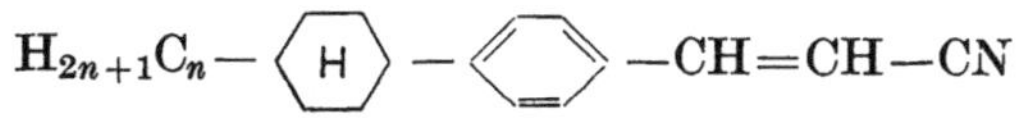 (1)

The compounds of the homologous series (1) differ from other binuclear liquid crystalline homologous series with the CN terminal group in that they reveal a smectic phase A for short alkyl chains, $n \geqq 4$ (DABROWSKI, SZCZUCIŃSKI).

## 2. Results and discussion

Table 1
Phase transition temperatures and molecular dimensions of EPCH

| $n$ | Cr | | $S_{Ad}$ | | N | | I | $l$ nm | $d_{ex}$ nm | $d_{calc}$ nm |
|---|---|---|---|---|---|---|---|---|---|---|
| 1 | 349 | — | | . | 229 | . | | | | |
| 2 | 322 | — | | . | 389 | . | | 1.79 | | |
| 3 | 323 | — | | . | 423 | . | | 1.94 | | |
| 4 | 330 | . | 305 | . | 417 | . | | 2.04 | | |
| 5 | 322 | . | 334 | . | 422.5 | . | | 2.19 | 2.87 | 2.82 |
| 6 | 327 | . | 379 | . | 417 | . | | 2.29 | 3.07 | 3.00 |
| 7 | 312 | . | 393 | . | 416.5 | . | | 2.43 | 3.27 | 3.30 |
| 8 | 321 | . | 401 | . | 410.5 | . | | 2.54 | 3.42 | 3.50 |

In Table 1 the phase transition temperatures are summarized of compounds (1) and the lengths ($l$) of their molecules in the most elongated conformation are compared with the spacing $d_{ex}$ of the smectic layers as measured from the scattering of X-rays by the method described by us earlier (DABROWSKI, PRZEDMOJSKI) and with those calculated from the composite structure of the dimer ($d_{calc}$). Compounds (1) revealing the smectic phase ($n \geqq 4$) have only one smectic phase which is of $A_d$ type since the spacing of the smectic layer is greater than the length of the single molecules (Tab. 1).

The ratio of the smectic layer spacing $d_{ex}$ to the length of the molecule ($l$) is almost constant, $d_{ex}/l \geqq 1.34 \pm 0.02$.

The structure of a dimer may be understood if the correlation between the experimentally observed spacing ($d_{ex}$) of the smectic layer and the length ($l_D$) of the dimer is established. The results of measurements of electric permittivity indicate that in the $A_d$ phase a state of equilibrium occurs between the concentrations of the monomer and dimer with antiparallel ordering of the dipole moments

$$D \rightleftarrows 2\,M\,.$$

The lowering of temperature increase the concentration of the dimer (LONGA et al.). This would indicate that the smectic layer spacing is determined by the equation

$$d_{ex} = x_M l_M + x_D l_D$$

where $x_M$ and $x_D$ are mole fractions of the monomer and dimer, respectively. X-ray investigations indicate that the spacing in smectics $A_d$ is independent of temperature or varies but slightly with temperature (HARDOUIN et al.; GUILLON et al.). Hence the growth of the dimer concentration does not cause any increase of the layer spacing. LONGA et al. have shown in terms of the microscopic model explaining the formation of the smectic $A_d$ phase and of the normal and reentrant nematic phases from terminally polar molecules that the forces induced between dimers, monomers, and between dimers and monomers under suitable conditions of temperature and density lead to a smectic layer structure with a period equal to the length of the dimer. On the basis of the above given premises we assumed that the length of the dimer determines the smectic layer spacing, $d_{exp} = l_D$, and we propose for the EPCH compounds the dimer structure shown in Figure 1.

In this model the degree of overlapping of the rigid cores into which the —CH=CH—CN groups have also been included depends on the assumed relation-ship between the length of the dimer and the spacing of the smectic layer and order parameter.

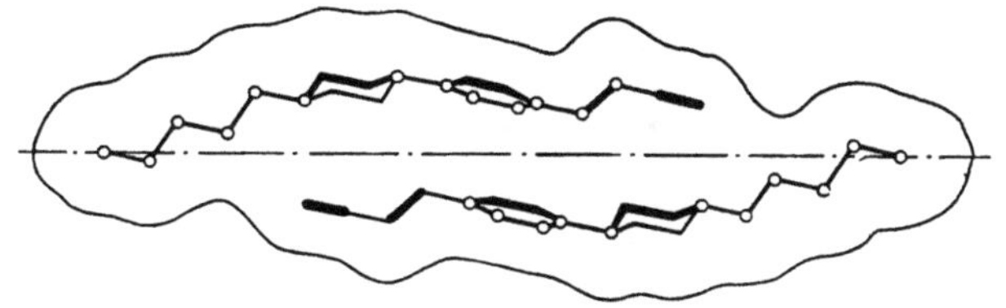

**Fig. 1.** Proposed structure of the dimer for EPCH compounds under the assumption that the dimer length is equal to the experimentally found smectic layer spacing, $d_{exp}$

If it is assumed that the spacing of the smectic layer is equal to the length of the dimer, then length of the region in which the rigid cores overlap is about 1.1 nm, i.e. the —CN group of one molecule lies above the centre of the cyclohexane ring of the other molecule (see Fig. 1). If we assume instead that this spacing is equal to the mean projection of the dimer length on the normal to the smectic layer, the ordering parameter being equal to 0.8, then the cores will overlap on the length of 0.8 nm.

In this case the CN group of one molecule lies over the bond connecting the phenyl ring with the cyclohexyl one. The problem of selection of one of the above models was discussed by us in an earlier paper (DĄBROWSKI, PRZEDMOJSKI).

In order to get a knowledge of the ability of the EPCH homologous series to create of the reentrant nematic phase we determine the phase diagrams by the single concentration method of the binary mixtures composed of a compound revealing the nematic phase only ($n = 3$) and of the first ($n = 5$) or the last $n = 8$ compound of the tested series having the enantiotropic smectic phase (see Fig. 2). The reentrant nematic phase is clearly visible in the 3EPCH–8EPCH binary mixture. In the

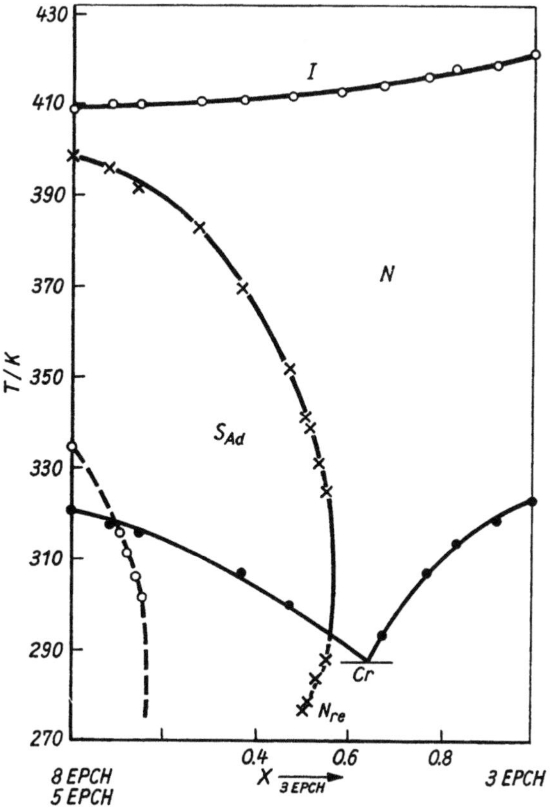


**Fig. 2.** Phase diagram of the 3EPC–H8EPCH binary system (solid line) on which a fragment of the phase diagram of 5EPCH–8EPCH system relating to the $S_{Ad} \rightarrow N$ phase transition (dashed line) has been plotted

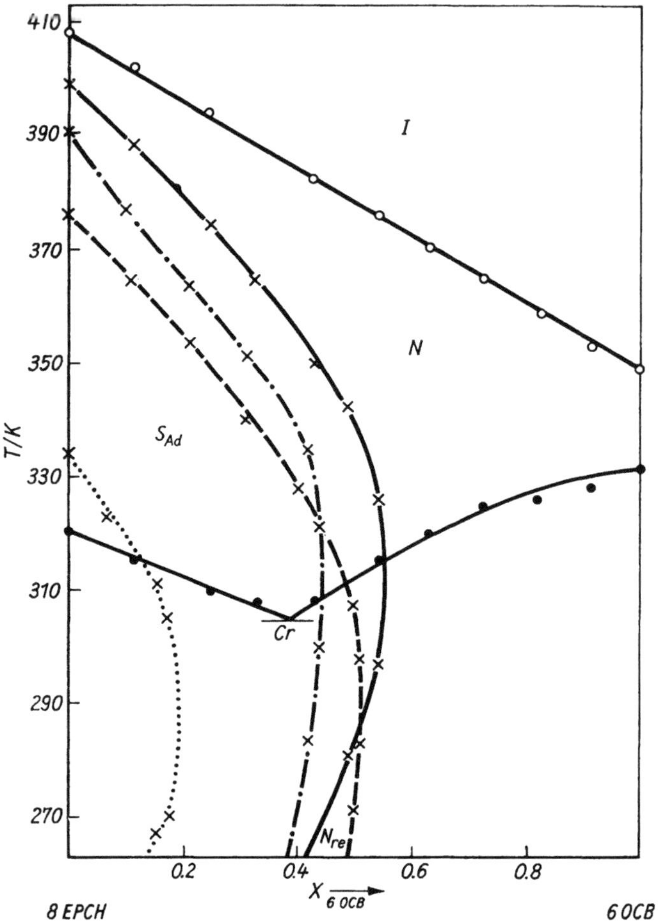


Fig. 3. Phase diagram of the 8EPCH–6OCB binary system (solid line) on which fragments of diagrams relating to the $S_{Ad} \rightarrow N$ phase transitions have been plotted for 5EPCH–6OCB (dotted line), 6EPCH – 6OCB (dashed line), 7EPC – H6OCB (dash-and-dot line)

3EPCH–5EPCH mixture the $A_d$ smectic phase is observed in a narrow concentration range, however, we did not succeed in supercooling these mixtures to temperatures at which the appearance of the $N_{re}$ phase could be expected.

On the other hand compound 5EPCH and all the successive members of this series reveal the $N_{re}$ phase in mixtures with 6OCB (see Fig. 3). The $N_{re}$ nematic phase is observed in all four tested mixtures, though in the 6OCB–6EPCH mixture it appears in the smallest concentration range (smallest bending of the $S_{Ad}$ phase boundary) and in the 6OCB–8EPCH mixture in the widest concentration range. In the latter mixture the $N_{re}$ phase is furthermore observed in the room temperature range what enabled us to determine the variation of the static electric permittivity (both its parallel, $\varepsilon_{||}$, and perpendicular, $\varepsilon_{\perp}$, component and of electric conductivity its parallel, $\sigma_{||}$, and perpendicular, $\sigma_{\perp}$, component) with temperature in three successive phases, N, $S_{Ad}$, and $N_{re}$. We selected a mixture of the mole fraction $x_{OCB} = 0.53$ and $x_{8EPCH} = 0.47$ characterized by the phase transitions: $N_{re}$ 290 $S_{Ad}$ 329 N 376 I. On transition from the nematic to the $A_d$ smectic phase we observe a small lowering of $\varepsilon_{||}$. This quantity assumes its

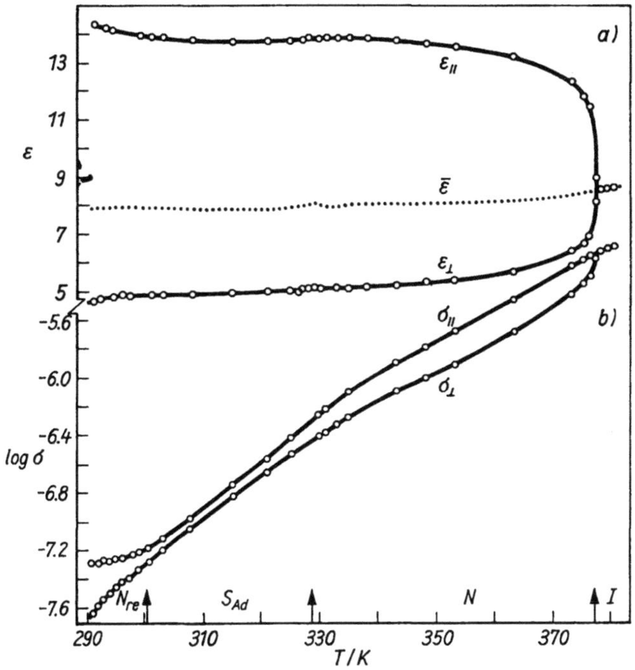


Fig. 4. Variation of the dielectric constants $\varepsilon_{\|}$ and $\varepsilon_{\perp}$ (a) and of the conductivity $\sigma_{\|}$ and $\sigma_{\perp}$ (b on temperature for the binary mixture of molar fraction concentration $x_{8EPCH} = 0.47$ and $x_{6OCB} = 0.53$ with a phase transition sequence $N_{re} \to S_{A_d} \to N \to I$

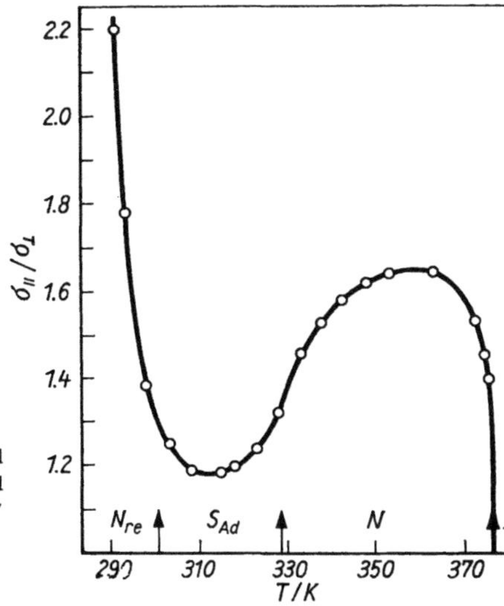


Fig. 5. Variation of the ratio of electric conductivity parallel to the long axis of molecules to the conductivity vertical to that axis in a mixture of composition $x_{8EPCH} = 0.47$ and $x_{6OCB} = 0.53$ revealing a reentrant nematic phase

smallest value at 313—318 °C upon which it increases more rapidly in the neighbourhood of the $A_d \to N_{re}$ transition and in the $N_{re}$ phase (see Fig. 4a).

The laminar structure of the smectic phase is strongly disturbed as it is seen from the X-ray patterns in which there are no weak higher order reflexes (STAMMATOFF et al). What is called a layer is usually only a small density modulation. It is in agreement with the observed changes of conductivity (see Fig. 4b and 5). In classic monolayer smectics (of $A_1$ type) the $N \rightarrow S_{A_1}$ transition is accompanied by a rapid decrease of conductivity in the direction parallel to the director, so that in the smectic phase we have $\sigma_\perp > \sigma_\parallel$ (HEPPKE and SCHNEIDER; MIRCEA-ROUSSEL et al.).

In the case of the investigated by us partially dimolecular smectic phase (of $A_d$ type) limited on both sides by the $N_{re}$ and N phase the changes of conductivity during the $N \rightarrow S_A$ transition are small. The relation $\sigma_\perp < \sigma_\parallel$ is preserved in the whole concentration range, though a distinct bothsided decrease of conductivity is observed which manifests itself by that the $\sigma_\parallel/\sigma_\perp$ ratio assumes the lowest value in the central region of the smectic $A_d$ range (see Fig. 5).

From the observed behaviour of conductivity of the smectic $A_d$ phase and the nematic reentrant and normal nematic phases we can conclude that the structure of the smectic $A_d$ phase is strongly perturbed, at least in the tested concentration range.

The $\varepsilon$ and $\sigma$ measurements have been made by means of the B 330 Wayne-Kerr bridge at 1.5 kHz. The condensator used was flat (1 mm layer) — the orienting field was 1 T.

Further details may be found in the work by JADŻYN and KĘDZIORA.

This work was supported by the Polish Academy of Sciences within the framework of Project CPBP 01.12.

## References

DĄBROWSKI, R., PRZEDMOJSKI, J., BARAN, J., Cryst. Res. Technol. **21**, 567 (1986)

DĄBROWSKI, R., SZCZUCIŃSKI, T., DZIADUSZEK, J.: Bul. Wojsk. Akad. Techn. No. 9, **33**, 27 (1984)

GUILLON, D., CLADIS, P. E., STAMATOFF, J.: Phys. Rev. Lett. **41**, 1598 (1978)

GÖBL-WUNSCH, A., HEPPKE, G., HOPF, R.: Z. Naturforsch. **36a**, 213 (1981)

HEPPKE, G., SCHNEIDER, F.: Ber. Bunsenges. Phys. Chem. **78**, 910 (1974)

HORDOUIN, F., LEVELUT, A. M., SIGAUD, G.: J. Physique **42**, 71 (1981)

JADŻYN, J., KĘDZIORA, P.: Mol. Cryst. Liq. Cryst., in press

LONGA, J., DE JEU, W. H.: Phys. Rev. **26A**, 1632 (1982)

MIRCEA-ROUSSEL, LEGER, L., RONDELEZ, F.: J. Physique C1, **36**, 93 (1975)

NGUYEN HUU TINH, J.: J. Chim Physique **80**, 82 (1983) and ref. cited there

STAMATOFF, J., CLADIS, P. E., GUILLON, D., CROSS, M. C., BILASH, T., FINN, P.: Phys. Rev. Lett. **44**, 1509 (1980).

*Authors' addresses*:

Prof. dr. R. DĄBROWSKI, B. SOSNOWSKA, doc. dr J. BARAN
Military Technical Academy
01-489 Warsaw, Poland

Doc. dr. J. JADŻYN
Institut Molecular Physics, Polish Academy of Sciences Poland,
Smoluchowskiego 17/19
60-179 Poznań

Doc. dr J. PRZEDMOJSKI
Institut of Physics, Warsaw Technical University Poland
Kyszykowa 75
00-622 Warsaw

Cryst. Res. Technol. | 22 | 1987 | 6 | K103–K104

**Short Notes**

G. A. Stergioudis, P. J. Rentzeperis, N. Sarafianos

Applied Physics Laboratory, Department of Physics,
Aristotle University of Thessaloniki, Greece, and
Laboratory of Physical Metallurgy, Department of Mechanical Engineering,
Aristotle University of Thessaloniki, Greece

# Crack Phenomena during Induced Crystallization of an Amorphous Alloy by Indentation

An amorphous alloy of stoichiometry $Cu_{60}Zr_{40}$ was crystallized by indentation methods at a temperature between $T_{gf}$ and $T_c$ (Stergioudis et al.).

After crystallization, crack attraction or repulsion due to the spherulitic crystalline phase dispersed in the bulk amorphous material, caused brittleness to the specimen. The phenomenon is more pronounced in the case of a specimen crystallized under vacuum conditions permitting controlled oxidation. This behaviour is not very different from that of a glass-ceramic (Chevalier et al.).

The ratio of the indentation crack length $C$ to the impression radius $\alpha$ is ver low in the single crystal phase (Fig. 1) and quite similar to that in other materials (Hockey,

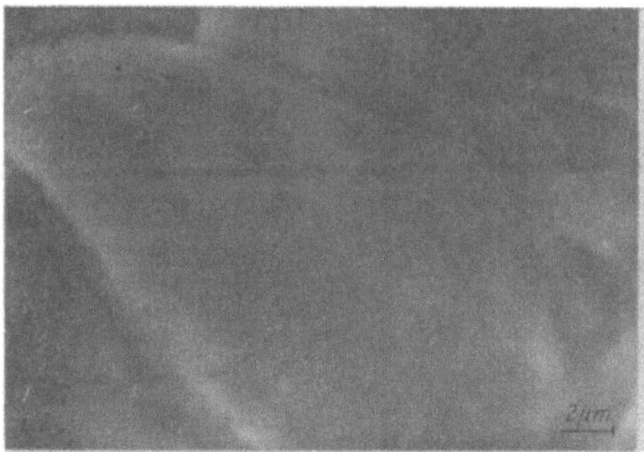

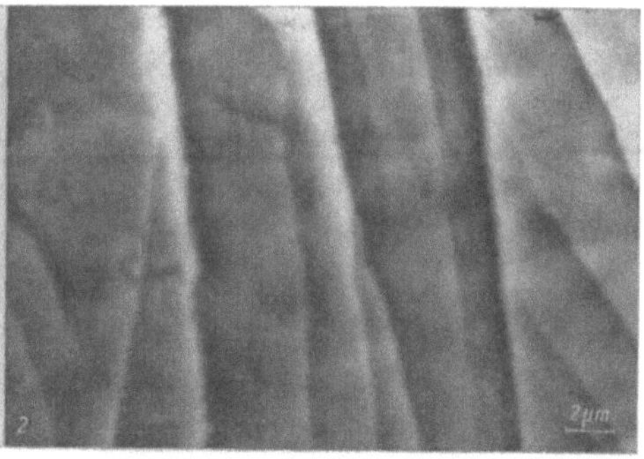

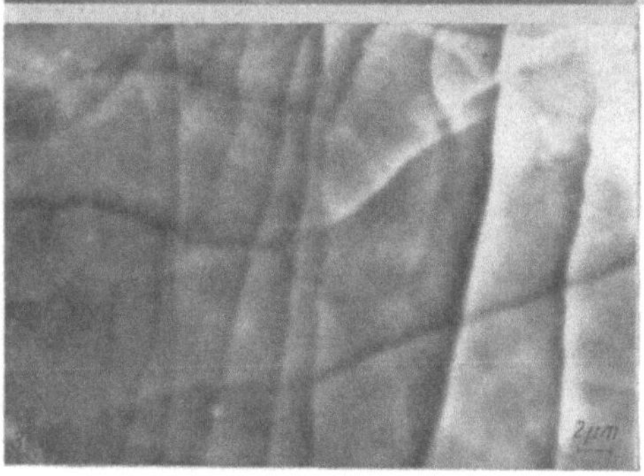

Fig. 1. S.E.M. showing cracks in an amorphous alloy of stoichiometry $Cu_{60}Zr_{40}$ crystallized to a single crystal phase by indentation

Fig. 2. Cracks and microcracks which appeared after the crystallization, was completed

Fig. 3. Extended microcracks overlapping main cracks when heat treatment is prolonged

LAWN). On the other hand, cracks are very numerous and intense when amorphous material transforms to polycrystalline.

Due to thermal expansion mismatch between crystalline and amorphous phase microcracks appear in addition to the main cracks (Fig. 2). Microcracking does not occur if the heat treatment is of a short duration. If the heat treatment is longer microcracking is evident, but slightly different from that observed by other investigators (FABER, EVANS). In our case microcracking extends to a large zone with frequent overlapping of cracks and microcracks (Fig. 3).

It seems that cracks are created when crystallization processes start and the external forces are still applied. When the experiments take place just above $T_{gf}$ crystals appear without cracking. This condition is difficult enough to achieve because of the shifting of temperature $T_c \rightarrow T_{gf} \rightarrow T_g$ during heating (GREST, COHEN; AVRAMOV et al.; BUSCHOW).

It seems that cracking depends on the heating rate, and it is more pronounced when the temperature rises quickly to reach $T_c$ than when the temperature rise is slower.

## References

AVRAMOV, I., GRANTSCHAROVA, E., GUTZOW, I.: J. Cryst. Growth **52**, 111 (1981)
BUSCHOW, K. H. J.: J. Appl. Physics **52** (5), 3319 (1981)
CHEVALIER, J. P., CALVAYRAC, Y., QUIVY, A., HARMELIN, M., BIGOT, J.: Acta Metallurg. **31** (3), 465 (1983)
FABER, K. T., EVANS, A. G.: Acta Metallurg. **31** (4), 557 (1983)
GREST, G. S., COHEN, M. H.: Phys. Rev. B, **21** (9), 4113 (1980)
HOCKEY, B. J., LAWN, B. R.: J. Mat. Science **10**, 1275 (1975)
STERGIOUDIS, G. A., JOURDAN, C., GASTALDI, J.: Mat. Res. Bull. **20**, 275 (1985)

(Received December 24, 1986)

*Authors' address*:

c/o Prof. Dr. P. J. RENTZEPERIS
Applied Physics Laboratory, Department of Physics
Aristotle University of Thessaloniki,
Thessaloniki, Greece

| Cryst. Res. Technol. | 22 | 1987 | 6 | K105—K108 |
|---|---|---|---|---|

J. KIRCHHOF, P. KLEINERT, A. FUNKE, B. KNAPPE, H.-R. MÜLLER

Physikalisch-Technisches Institut der Akademie der Wissenschaften der DDR, Jena

# About the Smoothing of Concentration Fluctuations by Diffusion in the Glass Systems $SiO_2/GeO_2$ and $SiO_2/GeO_2/B_2O_3$

Glass systems like $SiO_2/GeO_2/B_2O_3$ or $SiO_2/GeO_2/P_2O_5$ are the dominant materials in the fabrication of long fibers with low loss and wide band width as transmission medium for optical communication. Such fibers are drawn from preforms preferentially made by the so-called MCVD (Modified Chemical Vapour Deposition)-process (McCHESNEY et al.). This process includes the deposition of the above mentioned glasses within a fused quarz glass-tube by oxidation of gaseous halides at temperatures of 1700—2100 K and the subsequent collapsing of the tube into a solid rod. For making a fiber core with a graded refractive index profile about 30—60 layers with different compositions are deposited. The obtained result deviates from the desired ideally smooth index profile by a stair-like structure. Moreover, peculiar variations of the dopant concentrations and the refractive index in radial direction within a single layer appear, the reason of which is not fully clear until now (PRESBY et al.; AKAMATSU et al.). Such fluctuations lead to a visible ring structure in the cross-section of a preform core (Fig. 1) and may drastically increase the pulse dispersion and consequently reduce the transmission bandwidth (BEHM; COHEN et al.). It is obvious that the diffusion properties of the respective glass composition are of fundamental importance for a smoothing of the perturbed profiles and thereby for the improvement of the transmission properties of the fiber.

It is known that especially germanium as dopant exhibits no observable diffusion (FLEMING et al.). The fluctuations of concentration resulting from the deposition process are preserved through subsequent processing e.g. collapsing and drawing and finally appear in the fiber (PRESBY et al.; HOPLAND; BEHM; MARCUSE et al.). Only GUR'JANOV et al. have reported about a certain reduction of the fluctuations

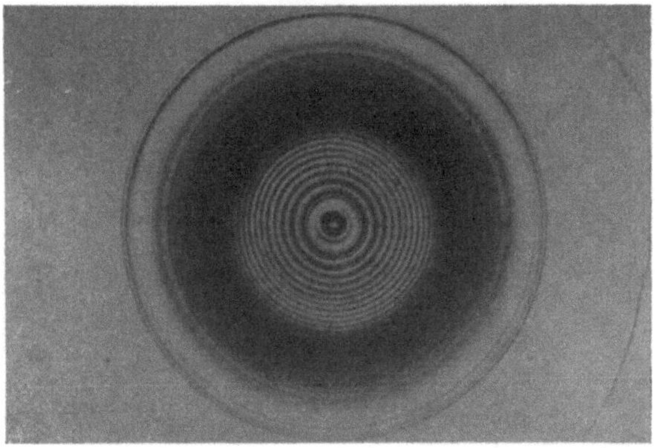

Fig. 1. Light microscope picture of a preform core

during the drawing process and Schneider et al. suppose a $GeO_2$ diffusion during the preform preparation to explain deviations from a sharply ascending step index profile. However, the defined conditions for a possible smoothing out of the $GeO_2$ fluctuations by diffusion are up to now fully unknown. Efforts aimed at a practicable technology for reducing the fluctuations were not successful (Behm).

In the following two preliminary experiments are described which give a clear proof for the appearance of a diffusive equalization of the spatial concentration differences in $SiO_2/GeO_2/B_2O_3$ and $SiO_2/GeO_2$ compositions.

In a first experiment a) after the deposition of four $SiO_2$ layers into a fused quartz glass tube (6 mm × 8 mm × 50 cm) one $SiO_2/B_2O_3$ layer with 7 mol% $B_2O_3$ and following six $SiO_2/B_2O_3/GeO_2$ layers with a constant composition of 7 mol% $B_2O_3$ and 14 mol% $GeO_2$ were deposited from a gaseous mixture of $SiCl_4$, $GeCl_4$, $BBr_3$, and $O_2$ flowing through the tube (layer thickness $\approx$10 μm). The tube was rotated in a glass working lathe and heated by an oxy-hydrogen torch, while moving along the tube with a speed of 10 cm $min^{-1}$. The temperature for the deposition of the doped layers measured with an infrared pyrometer at the outside of the tube was 2020 K in the maximum of the spatial profile. After the deposition a pressure of about 270 Pa was maintained within the tube for compensating the surface tension to prevent the collapsing (Kirchhof). The torch traverses the tube with a speed of 3 cm · $min^{-1}$, and the temperature was stepwise rised yielding zones of about 7 cm length treated at 2290 K (2240 K), 2350 K, 2400 K and 2450 K, respectively. Finally, the tube was collapsed to a solid rod by three passes of the torch with 10 cm · $min^{-1}$, 5 cm · $min^{-1}$ and 1.5 cm · $min^{-1}$, respectively, at a temperature of 2400 K without internal pressure. In a second experiment b) the same procedure was carried out but the deposited layers did not contain $B_2O_3$, and the deposition temperature was 2060 K.

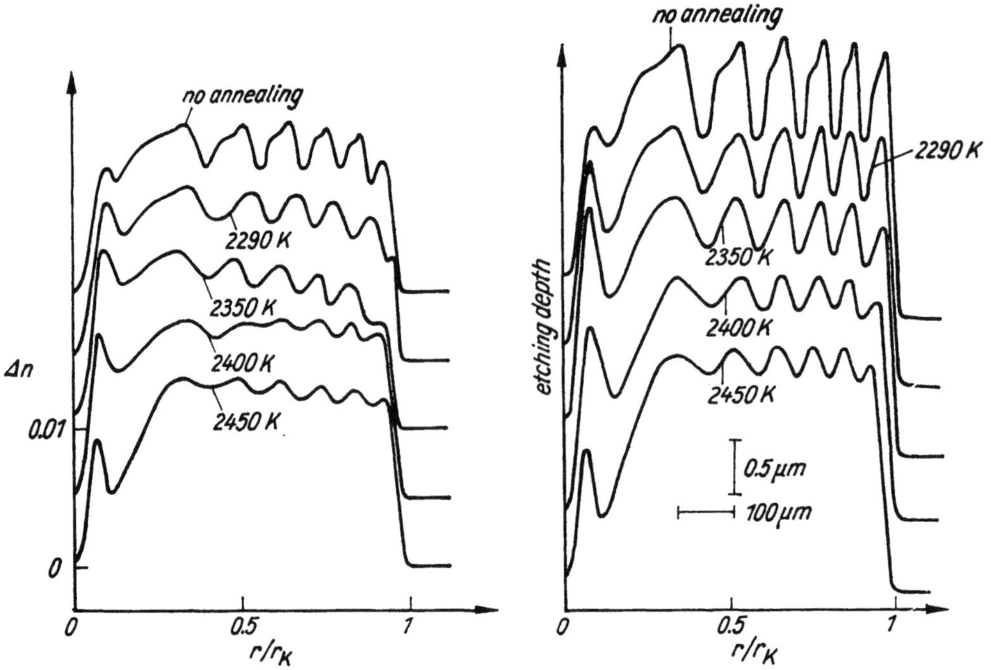


Fig. 2. Refractive index profiles and etching depth profiles of the preform core obtained at various annealing temperatures (Experiment b)

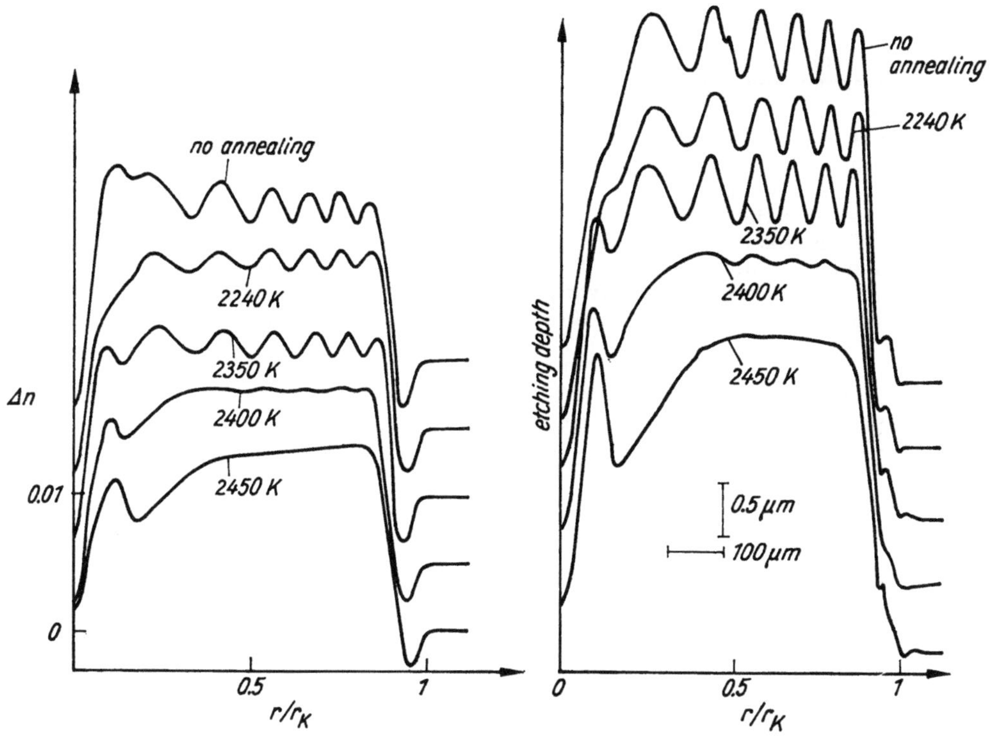


Fig. 3. Refractive index profiles and etching depth profiles of the preform core obtained at various annealing temperatures (Experiment a)

With a non-destructive technique (MÜLLER et al.) the refractive index profile of the core was measured at certain points corresponding to the above specified temperature treatment. Moreover, slices were cut from the rods and after a treatment with diluted hydrofluoric acid the etching profil was determined (KNAPPE, RÖPKE) with a stylus instrument (surface tester ME 10, ZEISS-Jena). The variations of the etching depth across the radius correspond to the fluctuations of the germanium content.

The results of the investigations clearly show the influence of the annealing procedure between deposition and collapsing on the distribution of the dopant concentration (FUNKE et al.). The fluctuations produced during the layer deposition begin to alter at 2300 K. Annealing at 2450 K leads to a considerable decrease for the $SiO_2/GeO_2$ system (Fig. 2). Obviously, the $B_2O_3$ content enhances the diffusion and leads at 2400 K to a nearly complete equalization (Fig. 3). Surprisingly, the collapsing at 2400 K without preceding temperature treatment seems to exhibit only a small effect, probably because the layer thickness and thereby the width of the fluctuations is quickly enlarged. An unequivocal determination of the diffusion during the collapse is not yet possible with the present experiments. It may be possible that by blowing up the tube during the annealing the temperature needed for the equalization can be further reduced as a result of the smaller diffusion length.

A rough estimation taking into account a diffusion length of $\approx 3 \cdot 10^{-4}$ cm and a diffusion time of $\approx 20$ s yields values of the order of $10^{-9}$ $cm^2 s^{-1}$ for the diffusion coefficients at 2400 K.

**References**

AKAMATSU, T., OKAMURA, K., UEDA, J.: Japan. J. applied Physics **16**, 1487 (1977)
BEHM, K.: AEÜ **32**, 403 (1978)
BEHM, K.: Dissertation, Braunschweig 1979
MCCHESNEY, J. B., O'CONNOR, P. B., DIMARCELLO, F. V., SIMPSON, J. R., LAZAY, P. D.: Proc. 10th Int. Congr. on Glass, Kyoto 1974, 6/40
COHEN, L. G., DIMARCELLO, F. V., FLEMING, J. W., FRENCH, W. G., SIMPSON, J. R., WEISZMANN, E.: Bell Syst. Techn. J. **57**, 1653 (1978)
FLEMING, J. W., KURKJIAN, CH. R., PAEK, U. CH.: J. Amer. Ceramic Soc. **68**, C-246 (1985)
FUNKE, A., KIRCHHOF, J., KLEINERT, P.: DDR-WP C 03 B 232 089, 18. 1. 1983
GUR'JANOV, A. N., DIANOV, E. M., LAVRIŠČEV, S. V., MAŠINSKIJ, V. M., NEUSTRUEV, V. B., NIKOLAJČIK, A. V., JUŠIN, A. S.: Kvantovaja Elektronika **6**, 2109 (1979)
HOPLAND, S.: Electronic Letters **14**, 757 (1978)
KIRCHHOF, J.: phys. stat. sol. (a) **60**, K127 (1980)
KNAPPE, B., RÖPKE, U.: 10th Conf. on Electronmicroscopy, Leipzig 1981, PO 1/91
MARCUSE, D., PRESBY, H. M.: Proc. IEEE **68**, 666 (1980)
MÜLLER, H.-R., RÖPKE, U.: phys. stat. sol. (a), to be published
PRESBY, H. M., STANDLEY, R. D., MACCHESNEY, J. B., O'CONNOR, P. B.: Bell Syst. Techn. J. **54**, 1681 (1975)
PRESBY, A. M., MARCUSE, D.: Proc. IEEE **68**, 1198 (1980)
SCHNEIDER, H., LEBETZKI, E., SCHIFFNER, A., WEIDINGER, F., AULICH, H., GRABMAIER, J. G.: Siemens Forsch.- u. Entwickl.-Ber. **9**, 53 (1980)

(Received December 24, 1986)

*Authors' address*:

Dr. J. KIRCHHOF, Dr. sc. nat. P. KLEINERT, Chem.-Ing. A. FUNKE, Dr. B. KNAPPE, Dr. sc. nat. H.-R. MÜLLER
Akademie der Wissenschaften der DDR, Physikalisch-Technisches Institut Jena
Helmholtzweg 4
Jena DDR-6900

| Cryst. Res. Technol. | 22 | 1987 | 6 | K109—K111 |
|---|---|---|---|---|

N. S. Peev

Institute of Microelectronic, Sofia, Bulgaria

# Morphology of SiC Epitaxial Layers Grown by Temperature Gradient Zone Melting (III)

## Polytypical Non-uniformity

### Introduction

It is underlined in the paper of Peev, et al. that the SiC epitaxial layers grown by temperature gradient zone melting in vacuum, are polytypical uniformity. Less than 10% of the grown layers contain other polytypical inclusions (usually polytype 4H) and the other 90% are polytypical uniformity, repeating the substrate's polytype 6H. Experiments are performed in order to change the grown polytype and to grow polytype together.

The growth process is described by Peev, et al.

### Discussion

In order to change the grown polytype the epitaxy is performed by solvent Tb–Si containing graphite powder. The participation of graphite leads to the appearance of polytype 4H–SiC together with the substrate's polytype 6H–SiC. 50% of the layers are the result of the growth of both polytypes and one of the experiments only 4H–SiC. It must be mentioned that the addition of graphite in the liquid phase leads to instability of the crystallization boundary, it bursts and bends significantly.

On some polytypical non-uniformity layers consisting of the unadvertently grown polytypes 4H and 6H the following investigations were performed. By thinning the layers and exposing them to ultraviolet light the 4H-region and the 6H-region are discovered and their areas measured by microscope. In Figure 1 the dependence of

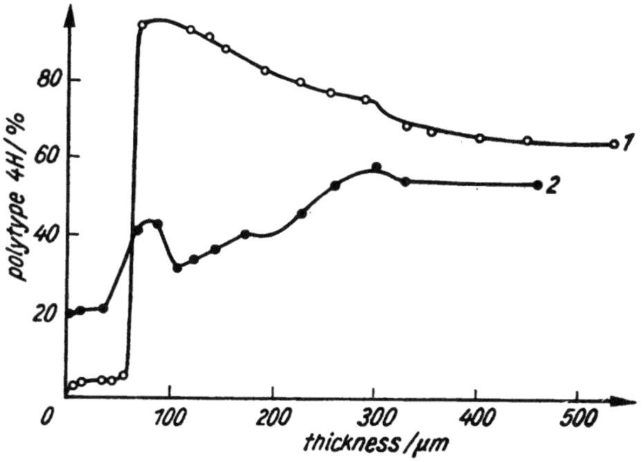


**Fig. 1.** Distribution of the polytype 4H-SiC along the thickness of the epitaxial layer expressed in percents of the area of this polytype: curve 1 — initial composition of the melt is 80 at.% Si – 20 at.% Tb; curve 2 — initial composition of the melt is 50 at.% Si – 50 at.% Tb

the area of polytype 4H (expressed in percents) on the layer thickness at initial compositions of the melt 80 at.% Si–20 at.% Tb (curve 1) and 50 at.% Si–50 at.% Tb (curve 2) is shown. In these experiments the substrates were polytypical uniformity of polytype 6H. From Figure 1 (curve 1) follows that the appearance of the new polytype proceeds monotonously, slowly and in 60 min (the growth rate is about 1 μm/min); it occupies only 5% of the whole surface, whereas in curve 2 the new polytype 4H–SiC occupies at once 20% of the surface. In both cases the arising of the new polytype takes place on the periphery of the layers: in this region the liquid phase is in direct contact with the environment. This circumstance indicates that the appearance of the new polytype is due to the influence of external factors.

The last 120 min both polytypes (4H and 6H) are in equilibrium, and it may be a thermodynamical equilibrium. This state of equilibrium takes place 320 min and 400 min, respectively, after the process begins; it is an equilibrium between two solid phases. The state of equilibrium in the case of curve 1 is characterized by the content of 65% polytype 4H and in case of curve 2 of 55%. This difference may be due to the influence of the liquid phase composition.

There exist five theories of polytypism (VERMA, KRISHNA) but only one of them is considered in this paper: the polymer theory which can be checked experimentally. For this purpose the following experiments were carried out: two "sandwich" structures were placed simultaneously on the graphite holder (Fig. 2). The substrate of the first structure was of the polytype 6H, and the seed crystal of the polytype 15R; whereas the substrate of the second structure was 15R and the seed crystal 6 H. The composition of the melts of both structures was identical. In this way in a growth process layers were grown from identical liquid phases, on different substrates and from different seed crystals. Taking into account that the substrate too is derived from the liquid phase, it may be concluded that the polymers 33 and 32 (or 23) in the zone are the polymers of the polytypes 6H- and 15R–SiC, respectively. In accordance with the polymer theory one may conclude that the possible polytypes are 33R, 51R-a, 69R, 87R, 21H, 78R, 27H, 84R. The precise polytype form will be determined by the ratio of the quantity of both polymers (VERMA, KRISHNA). The X-ray measurements have shown that only in two cases both polytype forms 6H–15R were grown together. In all other experiments the polytype form of the layer repeats the polytype form of the substrate. The polymer theory cannot explain these results.

Therefore the polymer theory is not useful for the liquid phase epitaxy because the influence of the substrate surface on the polytype form of the layer is not taken into account. It is more appropriate for gas phase epitaxy, for epitaxy of non-oriented substrates or in the case of crystal growth from three dimensional nucleations.

In result of all this the following conclusions are made:

— the kinetic processes on the substrate interface play an important role for the appearance of a new polytype form — accidental, significant or quick changes in the growth conditions lead to the appearance of a new polytype form,

— for the parallel growth of both polytype forms and for their reaching the state of equilibrium the thermodynamical factors are very important.

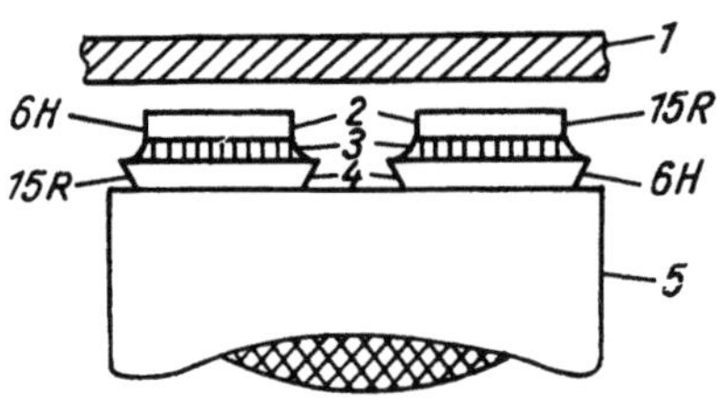


Fig. 2. Experiments to check- up of the polymer theory: 1 – graphite heater; 2 – seed crystal; 3 – liquid zone; 4 – substrate crystal; 5 – graphite holder

## Conclusion

In this way the parallel growth of both polytype forms takes place in two stages: 1. The appearance of the new polytype form, for which the kinetic factors are significant; 2. The simultaneous growth of both polytypes forms for which the thermodynamical factors are significant. It is known that the different polytype forms of SiC have approximately identical potentials (thermodynamical potentials). This conclusions can be made on the basis of the equilibrium state of the two polytypes (4H and 6H) in Figure 1.

## Summary

In the present paper, presented in three parts, the observed relationships in the morphology of the SiC epitaxial layers, grown by temperature gradient zone melting in vacuum are discussed. It is quite possible that some of the expressed notations are contraversial but they may be of interest to the specialist in epitaxy of semiconductors.

## References

PEEV, N. S., TAIROV, YU. M., SMIRNOVA, N. A., KALNIN, A. A.: Cryst. Res. Technol. **21**, 1503 (1986) (I), **21**, 1509 (1986) (II), **22**, 59 (1987) (III)

VERMA, A., KRISHNA, P.: Polymorphism i polytypism v crystallach, Moscow 1969

(Received November 4, 1986)

*Authors address:*

Dr. N. S. PEEV
Institute of Microelectronic
Boulv. "Lenin" 7 km
Sofia 1784
Bulgaria

# Crystal Research and Technology

## Zeitschrift für experimentelle und technische Kristallographie

Edited by H. NEELS, Leipzig

Papers in Number 7, 1987

### Original Papers

B. LORENZ: Grain Size Distribution in the Kolmogorov Model for Nucleation and Growth: Heterogenous and Homogeneous Nucleation

N. S. PEEV: Morphology of SiC Epitaxial Layers Grown by Temperature Gradient Zone Melting(II). The Dependence of the Structural Properties of the Epitaxial Layers on the Growth Conditions

H. KÜHNE: The Current Understanding of Epitaxial CVD Silicon Layer Doping in the Light of Modelling and Theory Development(IV). The Problem of More-than-one Kind of Dopant Species

R. FLAGMEYER, B. SCHUMANN: Characterization of Epitaxial Group II Fluoride Films on (100)-oriented GaAs Substrates by Ion Backscattering and Channeling

B. WOLF, A. ZEHE: Are In-grown Surface Reactions during MBE Controlled by a "Quasi-Gas Transition Layer"?

A. HUPFER, S. SCHULZE, D. HIRSCH, L. ŻDANOWICZ: Surface Preparation of $A^{II}_3B^V_2$ Semiconductors ($Cd_3As_2$, $Zn_3As_2$, $Cd_3P_2$, $Zn_3P_2$) under Ultra-high Vacuum Conditions

P. SEDLACEK, R.-J. KUBAN, K.-O. BACKHAUS: Structure Determination of Polytypes(II)

M. T. SEBASTIAN, P. KRISHNA: X-ray Investigation of the Mechanism of Phase Transformation in Single Crystals of ZnS, $Zn_xCd_{1-x}S$ and $Zn_xMn_{1-x}S$(I). Calculation of Diffraction Effects by a Three Parameter Model

A. WAŚKOWSKA, A. PIETRASZKO, Z. CZAPALA: Optical, X-ray, and Calorimetric Studies of Phase Transitions in $ND_4BeF_3$

J. HÄRTWIG, K.-H. JÄCKEL, V. LERCHE: Simulation of Contrasts in X-ray Plane Wave Topographs of Quartz Crystals with Induced Growth Striations

G. BOCHMANN, E. KLAUS, G. FRITSCHE, W. WAGNER: Phase Structure of Boride Layers on Iron Substrates with Electroplated Cobalt Layers

K. D. GLINCHUK, V. I. GUROSHEV, A. V. PROKHOROVICH: Irradiation Plus Annealing-induced 1.20 eV Emission Band in p-Type GaAs

P. VENUGOPAL REDDY, K. J. PRATAP, T. SESHAGIRI RAO: Electrical Conductivity of Some Mixed Ferrites at Curie Point

J. W. TOMM, B. SUMPF, Karin HERRMANN, A. SZERBAKOW: Some Aspects of the Technology of Lead Salt Diode Lasers Used in Gas Monitoring Systems

AKADEMIE-VERLAG · BERLIN

# Systems Analysis – Modelling – Simulation

Journal of Mathematical Modelling and Simulation in Systems Analysis

Edited at the Central institute of Cybernetics and Information Processes of the Academy of Sciences of the G.D.R. in cooperation with the Mathematical Society of the G.D.R. by Achim Sydow (Berlin) with the assistance of an international team of scientists

Published yearly, 1 volume (6 numbers).
Volume 3, 1987.
16,7 cm by 24,0 cm. Annual list of contents.

Annual subscription M 120.—; single number DM 30.—.
Order number: 1097 ISSN 0232-9298


The journal is devoted to art and techniques and applications of modelling and simulation in systems analysis. Systems analysis is an interdisciplinary activity. Articles will be published on systems analysis and systems theory, modelling and mathematical models and tools for decision and control, computer simulation systems, modelling and simulation in environmental protection, macroeconomics, agricultural production, regional planning, biosciences, traffic control, large engineering systems, resources papers, distribution, management systems etc. In addition to original survey articles and reports on conferences etc. are published. Finally, books covering the above mentioned and related subjects are reviewed.
Of interest to: Scientists of systems sciences, informatics, control engineering, applied mathematics, operation research, biology, economy, physics, environmental protection, agriculture etc.

Orders can be sent

— in the GDR: to the Zeitungsvertriebsamt der Deutschen Post, Abt. Zentralvertrieb Berlin, Straße der Pariser Kommune 3—4; 1004 Berlin, or to the Akademie-Verlag Berlin, Leipziger Str. 3—4, PF-Nr. 1233, DDR-1086 Berlin,

— in the other socialist countries: to a bookshop for foreign language literature or to the competent news-distributing agency;

— in the FRG and Berlin (West): to the wholesale distributing agency Kunst und Wissen, Erich Bieber OHG, Wilhelmstr. 4—6, D-7000 Stuttgart 1;

— in the other Western European Countries: to Kunst und Wissen, Erich Bieber GmbH, General Wille-Str. 4, CH-8002 Zürich;

— in other countries: to the Akademie-Verlag Berlin, Leipziger Str. 3—4, PF-Nr. 1233; DDR-1086 Berlin.


AKADEMIE-VERLAG BERLIN
THEORIA CUM PRAXI
Deutsche Demokratische Republik
DDR-1086 Berlin, Leipziger Strasse 3-4 Postfach 1233

**Crystal Research and Technology**
Volume 22 1987 Number 6

# Contents

**Original Papers**

B. BŘEZINA, M. HAVRÁNKOVÁ: Crystal Growth and Domain Structure of Triglycine Sulphate Doped with Pt(II) . . . . . . . . . . 753
V. TUČEK, S. HOLEČEK, L. KUBÍČEK: Natural Convection during Directional Solidification of an Al-Zn Eutectic . . . . . . . . . . 761
H. KÜHNE: The Current Understanding of Epitaxial CVD Silicon Doping in the Light of Modelling and Theory Development(I). The Law of Epitaxial Layer Doping and its Limitation . . . . . . . . . . 767
Ch. GEORGI: On the Origin of the (1$\bar{2}$16)-Orientation in Epitaxial AlN Layers on R-Plane Sapphire . . . . . . . . . . 777
N. S. PEEV, Yu. M. TAIROV, N. A. SMIRNOVA, A. A. KALNIN: Morphology of SiC Epitaxial Layers Grown by Temperature Gradient Zone Melting(I). Dependence of the Crystallization Boundary Smoothness on the Epitaxial Layers Growth Conditions . . . . . . . . . . 783
K. SANGWAL: A New Equation for the Temperature Dependence of Density of Saturated Aqueous Solutions of Electrolytes . . . . . . . . . . 789
P. SEDLACEK, R.-J. KUBAN, K.-O. BACKHAUS: Structure Determination of Polytypes(I) . . . . . . . . . . 793
R.-J. KUBAN, B. SCHULZ: Crystal Structure Determination of 2-Acetyl-amino-4-benzoyl-5,5-dimethyl-4,5-dihydro-1,3,4-thiadiazolin, $C_{13}H_{15}N_3O_2S$ . . . 799
C. C. DESAI, A. N. HANCHINAL: Microstructures of Gel-grown Ammonium Hydrogen Tartrate Single Crystals . . . . . . . . . . 803
C. C. DESAI, M. S. V. RAMANA: Electrical Conduction in Ferroelectric Lead Hydrogen Phosphate Single Crystals . . . . . . . . . . 809
G. HAUCK: Surface Structure of Obliquely Deposited Layers and Liquid Crystal Alignment . . . . . . . . . . 817
K. G. SUBHADRA, K. A. HUSSAIN: Melting Temperature of $NaClO_3$–$NaBrO_3$ Mixed Crystals . . . . . . . . . . 827
Z. RASZEWSKI, R. DĄBROWSKI, Z. STOLARZOWA, J. ZMIJA: Dielectric Studies on Binary Mixtures Containing 4-Trans-4'-n-hexyl-cyclohexyl-isothiocyanato-benzene . . . . . . . . . . 835
B. WEINERT, P. KIRSTEN, W. SIEGEL: The Influence of Crucible Moving on Growth Rate and Perfection of SSD-GaP Crystals . . . . . . . . . . 845
Sr. PETROV, Iv. CHOMAKOV, B. I. BOGDANOV: X-ray Quantitative Determination of the Phases in a Batch for the Synthesis of Poly- and Monocrystal Yttrium aluminium Garnet . . . . . . . . . . 855
R. DĄBROWSKI, J. JADŻYN, J. PRZEDMOJSKI, J. BARAN: The Nematic Reentrant Phase in Binary Mixtures Composed of a 4-(trans-4-Alkylcyclohexyl)-1-(2-Cyanoethenyl)-benzene and 6OCB . . . . . . . . . . 861

*Continued on cover page 3*

Crystal Research and Technology is indexed in Current Contents/PC & ES.

*Continued from cover page 4*

**Short Notes**

G. A. Stergioudis, P. J. Rentzeperis, N. Sarafianos: Crack Phenomena during Induced Crystallization of an Amorphous Alloy by Indentation . . . K103

J. Kirchhof, P. Kleinert, A. Funke, B. Knappe, H.-R. Müller: About the Smoothing of Concentration Fluctuations by Diffusion in the Glass Systems $SiO_2/GeO_2$ and $SiO_2/GeO_2/B_2O_3$ . . . . . . . . . . . . . . K105

N. S. Peev: Morphology of SiC Epitaxial Layers Grown by Temperature Gradient Zone Melting (III). Polytypical Non-uniformity . . . . . . . . . . . K109

**Book Reviews**

U. Gonser: Microscopic Methods in Metals (O. Brümmer) . . . . . . . . . . . . 782

V. D. Scott, G. Love (Eds): Quantitative Electron-Probe Microanalysis (M. Wendt) 788

J. S. Blakemore: Solid State Physics (P. Rennert) . . . . . . . . . . . . . . 808

Manfred Grasserbauer, Hans Joachim Dudek, Maria F. Ebel: Angewandte Oberflächenanalyse mit SIMS, AES und XPS (O. Brümmer) . . . . . . . 816

R. Glocker: Materialprüfung mit Röntgenstrahlung. Unter besonderer Berücksichtigung der Röntgenmetallkunde (P. Paufler) . . . . . . . . . . 826

Bei Fragen zur Produktsicherheit wenden Sie sich bitte an:
If you have any questions regarding product safety,
please contact:

Walter de Gruyter GmbH
Genthiner Straße 13
10785 Berlin
productsafety@degruyterbrill.com